Cro-Magnon

Cro-Magnon

The Story of the
Last Ice Age
People of Europe

Trenton Holliday

Columbia University Press

New York

Columbia University Press
Publishers Since 1893
New York   Chichester, West Sussex
cup.columbia.edu

Library of Congress Cataloging-in-Publication Data
Names: Holliday, Trenton, author.
Title: Cro-magnon : the story of the last Ice Age people of Europe / Trenton Holliday.
Description: New York : Columbia University Press, 2023. |
Includes bibliographical references and index.
Identifiers: LCCN 2022054947 (print) | LCCN 2022054948 (ebook) |
ISBN 9780231204965 (hardback) | ISBN 9780231204972 (trade paperback) |
ISBN 9780231555777 (ebook)
Subjects: LCSH: Cro-Magnons. | Paleoanthropology. | Glacial epoch. |
Human beings—Effect of climate on.
Classification: LCC GN286.3 .H65 2023 (print) | LCC GN286.3 (ebook) |
DDC 569.9—dc23/eng/20230113
LC record available at https://lccn.loc.gov/2022054947
LC ebook record available at https://lccn.loc.gov/2022054948

Printed and bound by CPI Group (UK) Ltd, Croydon, CR0 4YY

Cover design: Julia Kushnirsky
Cover photograph: Alamy

For Kathleen

Contents

Contents

Preface

Lar and Mil were dizzy with adrenaline, their palms slick with sweat. Today they were taking part in their first reindeer hunt. For as long as they could remember, they had been honing their skills throwing spears; thirteen-year-old Mil threw spears with deadly accuracy, and twin Lar was not far behind. Silent and nervous in their blind, they waited for the signal to attack—a whistle from one of their cousins. They remained motionless in the cold autumn air so as not to scare away the approaching herd.

The year had been a tough one. Last fall the reindeer migration had been meager; the group had killed only eighteen, and of those, three were judged too sick to eat. The old ones said they had never seen such a poor hunt. With forty-four mouths to feed, fifteen reindeer would not sustain them over the long winter. After smoking the meat and preparing the skins, the six families held an emergency council. It was decided that the best course of action would be to head far to the south to harvest hazelnuts. Grandma Ela was from the south and spoke three southern languages, so she would be their guide for the two-week journey.

Along the way the mothers set snares for rabbits and hares every evening. Rarely were these efforts unrewarded, but a rabbit or two a day per family, along with a bit of cured reindeer meat, was less sustenance than they normally had. Fortunately, when they reached their destination, the hazelnuts were plentiful, and everyone—young and old—helped crack them open.

They overwintered in a sheltered valley Grandma Ela knew from her youth. She led them to an impressive rock shelter in a wooded glen, and thankfully it was unoccupied. However, it was clear that people had lived there in the past—black charcoal circles told of fires tended long ago, and discarded stone tools and chips of stone from tool making were scattered here and there. Grandma Ela picked up an old flint blade and said she was certain she had made it when she was young; it looked like her handiwork. Lar wasn't sure she was serious. Grandma, with her funny accent, incorrect grammar, constant jokes, and strange sayings, was considered by all to be hilarious—and she was always teasing Lar.

Winter was a lean time of long nights and short days, and they subsisted for the most part on hazelnuts, small amounts of cured reindeer meat, and the occasional rabbit. On a particularly frigid but clear day, their cousin Gulic killed a red deer, and everyone shared in the bounty. Feast days like these were few and far between, and for much of that winter food was carefully rationed and the kids' bellies ached. But at nights around the fire they would tell the old stories and sing the old songs, and Aunt Ulen would play her flute. Yes, food was scarce, but it felt like home.

One evening four strangers appeared at the shelter. Grandma talked to them, and Mil and Lar eavesdropped because Grandma had taught them her language. The strangers were impressed with Grandma's lineage. Grandma Ela was descended from a tree goddess? Why had she never mentioned this? Did that mean that Mil and Lar were too? They stared at each other with open mouths. Grandma then told a joke, and the tension in the strangers' faces dissolved away as they laughed heartily. It was clear that it wasn't just northerners who found Grandma hilarious!

The next day the strangers returned with sausages. These were cooked and shared with all. Grandma acted as an interpreter between the groups, and Lar and Mil showed off a bit by speaking to the strangers in their own tongue. The next morning, before departing for good, the strangers wished everyone well and embraced Grandma for a long time before leaving. Grandma tried to hide her tears, but Mil saw them.

With the arrival of spring, plant foods became more available. For the kids this was both a blessing and a curse. The worst food they ate that spring, at least in Mil's and Lar's opinion, had been cattail root cakes made from flour that the two had been forced to grind themselves. They made a secret pact that as adults they would never eat cattail roots again.

Summer meant it was time to return to the homelands. Within a week they were back in familiar surroundings, and the kids began to recognize the locations where they had camped. Food remained plentiful; no one went hungry. Near summer's end, the children joyfully ate blackcurrant berries until their hands were stained a deep purple. The shortening of the days and the first chill wind told the families that now was the time to make their way to the fall hunting grounds. Everyone was hopeful that this year the reindeer would come.

And come they did. Gulic's mother raced into camp from her lookout post high on a nearby ridge. The fall's first herd of reindeer, much more numerous than last year, was headed to the choke point in the ravine. Everyone sprang into action. Mil and Lar grabbed their spear throwers and darts and joined the able-bodied adults in the first party to leave camp. The grandparents and young children would follow later, for butchering more than twenty reindeer would require everyone's help.

From their blind, Mil and Lar could see that the herd was now climbing up the narrow ravine—for miles this was the only route up the cliffs and out of the broad valley where the deer had spent the summer eating to their hearts' content. "You must let the first ones go by before you throw; once they've gotten up to where your cousins are hiding, then we attack," their father had instructed them. It seemed to Lar like the whole herd was going to pass without anyone throwing a spear. Then the whistle—loud and piercing—rang out. Lar and Mil leapt up and let their first spears fly. Both struck reindeer, penetrating deep into their sides. They loaded another dart and then another, as did all the hunters. For several minutes it seemed as if the air was filled with spears. Because the deer were so tightly packed, most darts hit their targets. This year there would be plenty of food to last the winter.

This story took place 32,000 years ago. Due to the rise in sea level that began more than 15,000 years ago, the fall hunting grounds of Mil's and Lar's people now lie 40 meters (130 feet) beneath the English Channel. Their group's emergency refuge was a rock shelter near what is now Poitiers, France, some 430 kilometers (267 miles) to the south. Today Lar's and Mil's people are colloquially known as Cro-Magnons—people physically indistinguishable from you and me who made their living hunting and gathering in Ice Age Europe. These were people who loved, cried, and laughed, and like you and me they had their own struggles and triumphs. They made music, painted the walls of caves and huts, and created sculptures from clay, rock, ivory, and antler. They buried their dead, but they left no written records. All we know about them we have learned from studying their skeletons and the art and debris they left behind as they went about their lives. Although the story of Lar and Mil is a product of my imagination, no doubt there were people in Ice Age Europe who lived lives much like those depicted here. This book tells their story. It includes vignettes like this one, but it's primarily about the scientific investigation of these ancient people, written by someone who has studied them for more than three decades. I find the Cro-Magnons fascinating, and my hope is that by the end of this book you will as well.

Cro-Magnon

Introduction

SETTING THE STAGE

A little over 150 years ago, railroad workers in the picturesque village of Les Eyzies-de-Tayac-Sireuil in the Vézère Valley of the Périgord region of southwest France accidentally discovered multiple human skeletons associated with Paleolithic ("Old Stone Age") tools and ancient fauna, including mammoth and reindeer.[1] The remains were found in a small rock shelter known as *l'abri Cro-Magnon*. "Abri" is French for "shelter" or rock shelter,[2] and "cro" is an alternative spelling of the Occitan[3] word *cròs* or *clòt*—a cavity or hollow (*creux* in French). *Magnon* was the property owner's name; the property was listed as "Cramagnon" in the public registry, according to Lartet.[4] As the first verifiable discovery of *Homo sapiens* skeletons found with extinct animals, these finds were so noteworthy that even today the term "Cro-Magnon" is widely understood to represent a form of prehistoric human, despite the fact that this appellation has almost completely disappeared from the scientific literature. Nonetheless, Cro-Magnon was for many years a widely used name for the earliest modern humans in Europe. For example, in the 1980s, fossils from much earlier contexts in the Levant (the eastern shore of the Mediterranean) were sometimes called "Proto-Cro-Magnons" (chapter 4).

To the layperson, it may sound odd to refer to prehistoric people from the Cro-Magnon rock shelter—folks who lived about 32,000 years ago and who used stone tools—as "modern." However, in the field of human paleontology, modern humans are not "modern" in the sense that modern art, modern architecture, or modern dance is modern. In fact, fossils of modern, or nearly modern, humans can be quite ancient, dating back to perhaps 300,000 years ago in northwest Africa (chapter 4). The skeletons from the site of Cro-Magnon are modern in that they are members of our own evolutionarily successful and cosmopolitan species, *Homo sapiens*. In human paleontology, the Linnaean name *Homo sapiens* and the more vernacular terms "modern humans" or "anatomically modern humans" are often synonymous.[5]

As for *H. sapiens*, we are the only surviving species of a long and fairly bushy lineage of African primates known as hominins, members of the taxonomic tribe Hominini. In Linnaean taxonomy, a tribe is more inclusive than a genus but less inclusive than a subfamily. Among our living relatives, chimpanzees, bonobos ("pygmy chimpanzees"), and humans are all members of the subfamily Homininae. At the tribe level, however, chimpanzees and bonobos are members of the tribe Panini, a name that makes me think of Italian bread, and not members of our tribe, Hominini. Hominins are therefore quite simply defined as those animals more closely related to us than to chimpanzees or bonobos.

Although there is no agreement among scientists as to their exact number, today more than twenty-five fossil hominin species are reported in the scientific literature. Some are almost certainly our ancestors but most are collateral relatives—cousins in hominin lineages that either went extinct or taxa who despite no longer being with us nonetheless contributed a small number of genes to our more direct ancestors—genes still found in people today. In chapter 6, you will meet one 40,000-year-old early modern human from Romania who may have had a great-great grandparent who was a Neandertal (*H. neanderthalensis*, sister species of *H. sapiens*).[6]

This book is dedicated to telling the story of the Cro-Magnons from two perspectives. The first is that of a biological anthropologist, the field in which I am trained. Biological anthropologists (sometimes abbreviated as bioanthropologists) study the evolution, behavior, and genetic and morphological variation of humans, our living primate relatives (lemurs, lorises, tarsiers, monkeys, apes), and the fossil remains of our hominin ancestors and their relatives.

The second perspective is that derived from Paleolithic archaeology, an area in which I have much practical experience but less formal training. Archaeologists study the debris left behind by prehistoric people—from the remains of their living spaces (caves, huts, or other temporary shelters; postholes, internal structures, hearths) to the food they ate (remains and residues of animals and plants, sometimes burned), to the tools they used (typically stone because it tends to survive due to its durability, but sometimes wood, bone, or antler), to their funerary practices (intentional burials and associated grave goods). At least for later human evolution, which here I equate with the beginning of the Late Pleistocene about 129,000 years ago,[7] most fossil hominin skeletons have been recovered by archaeologists working in controlled excavations of hominin sites.

I am also a paleoanthropologist. Paleoanthropology is the multidisciplinary study of human evolution, and the bulk of paleoanthropologists are either human paleontologists (bioanthropologists like me who study fossil humans) or Paleolithic archaeologists, who study the materials these humans left behind in the distant past. Both of these scientific approaches are critical to understanding the evolution and behavior of prehistoric people, but given my training, the primary approach this book takes is one of biological anthropology. From this perspective, although we acknowledge that everyone alive today is a member of *H. sapiens*, when it comes to identifying fossil members of our species, things are murkier. The overall gestalt of the Cro-Magnons is one of anatomical modernity; however, in some aspects of their anatomy, they do not always fall within the range of variation seen among people today. Should it surprise us that people 30,000 years ago do not look exactly like we do now? This begs the question— who are the earliest *H. sapiens*, and how do we identify them (chapter 4)?

I also investigate the evolutionary origins of *H. sapiens*, the role of Neandertals therein, and touch on the continuing debate surrounding modern human behavior—whether such a thing exists, and if it does, whether it can be limited to our species alone. This is followed by discussion of the archaeological and paleontological evidence of the Cro-Magnons—including their biology, food, technology, and art—from their earliest known appearance in Europe about 54,000 years ago through the end of the Pleistocene about 11,700 years ago. I finish with a discussion of how the Cro-Magnons dealt with climate change—both when climate became much colder than today and when temperatures rose, glaciers retreated, and Europe became reforested. How they responded to climate change may offer valuable lessons for us today.

Who am I? I have been a fossil hunter for nearly as long as I can remember. As a child, I spent hours scouring gravel driveways for marine fossils. Crinoids, marine animals also known as sea lilies, were among my favorite discoveries. I was a quirky kid; another of my favorite pastimes was to randomly pull out a volume of the encyclopedia to read. I first read about human evolution in a *World Book* encyclopedia article written by my "academic grandfather" (the doctoral advisor of my doctoral advisor), the paleoanthropologist Alan Mann. But I had already been captivated by the subject at age eight, when my mother told me about the discovery of "Lucy"—a 40 percent complete, 3.18-million-year-old *Australopithecus afarensis* skeleton—at the time our earliest human ancestor. As a teenager, this burgeoning interest in human origins led me to study anthropology in college. It was there I became enamored with later human evolution, especially modern human origins. The origin of modern humans is the oldest question in paleoanthropology: How and through what evolutionary processes did people like you and me come to exist? It was the topic of my doctoral dissertation, and this question continues to enthrall me. In this book I detail how fossil and genetic evidence together weave threads of a fascinating and complicated story of our own species' evolutionary emergence (chapters 5 and 6).

Throughout this book are vignettes from my time researching fossils; through these I hope to give you a feel for what it's like to be a member of the cadre of people who study human evolution. It may not surprise you to learn that I'm a quirky adult—after all, one must be a bit eccentric to study human evolution for a living. Most of the time I am blissfully unaware of this, but there are times when it's painfully obvious. For example, strolling across campus one day, I saw a student walking toward me wearing a black T-shirt featuring a beautifully rendered skull in front of a satellite photo of Earth. I found myself assessing the taxonomy of the skull on his shirt: Was it a *Homo erectus*? A Neandertal? Up close, I saw that his shirt read "Recycle or Die." Should I be disturbed that for me skulls are objects of study that no longer represent danger or death?

EXCITING TIMES

Before we begin the Cro-Magnons' story, some historical background is warranted. Europe in the first half of the nineteenth century was in the throes of a scientific revolution, with great advances being made in geology and biology—advances that were changing our perception of the planet and our place in the universe. For centuries in Europe, Earth and the animal and plant species inhabiting it had been viewed as static and unchanging. Earth was created by God in its current state about 6,000 years ago. No new species had emerged; none had gone extinct. The notion that there was such a thing as prehistoric humans was considered fallacious. We knew what had happened since the earliest days of creation because it had been written down in the books of what Christians, by far the predominant religion in Europe, refer to as the Old Testament. There was no such thing as prehistory; all human existence had been recorded, and the world was young and static. In the lyrics of the *Gloria Patri*, "As it was in the beginning, is now and ever shall be, world without end"—yet this view of a young and static Earth was to undergo a radical revision.

In Earth sciences, the Scottish geologist Charles Lyell (1797–1875), following in the footsteps of his countryman James Hutton (1726–1797), embraced and further refined the scientific framework of uniformitarianism, the idea that the same geological processes that affect the Earth's surface today—volcanism, wind and water erosion, frost weathering, soil deposition, glaciers, earthquakes—operated to shape the Earth's crust in the past. Unlike the competing model of catastrophism, which viewed most features of Earth's crust as the result of cataclysmic events, uniformitarianism implied a much older age for the Earth. If the Grand Canyon in what was then called the New Mexico Territory had not been born of a single earth-rending event, but rather had been steadily eroded away by the tiny Colorado River at its base, then the Earth was very ancient indeed, certainly older than 6,000 years! Lyell's *Principles of Geology* was a veritable *tour de force* of the uniformitarian perspective and led to the widespread abandonment of catastrophism.[8] As a result, scientists in the mid-nineteenth century were beginning to accept a much older age for the Earth than had previously been thought possible. It is Lyell to whom we owe the broad acceptance of the concept of deep (geological) time.

The acceptance of an ancient Earth is important because deep time is essential for evolutionary biology, as broadly speaking it is a necessary condition for macroevolution (i.e., the origin of new taxa or the bifurcating speciation events leading to the formation of the "tree of life") to occur. As such, Charles Darwin's (1809–1882) formulation of the origin of species via natural selection depended heavily on deep time. An avid pigeon breeder, Darwin was aware of the number of varieties of, and amount of morphological change humans had wrought on, the rock dove (*Columba livia*) in a relatively brief amount of time. The catch is that these varieties are still pigeons—they remain members of the same species[9]—and therefore the differences manifest between them are classified as microevolutionary rather than macroevolutionary in character. The prevailing view among nineteenth-century naturalists was that varieties were ontologically[10] different from species, but Darwin took the contrarian position. For him, varieties that had emerged in human time had the potential to become differentiated species given enough geological time. Species were "only strongly-marked and well-defined varieties."[11]

Although Darwin is rightly considered to be the preeminent evolutionary scientist of all time, the truth is that early nineteenth-century Europe was awash with evolutionary ideas. Darwin's own grandfather, Erasmus Darwin (1731–1802), had been an evolutionist of some renown. The French naturalist Jean Baptiste de Lamarck (1744–1829), although criticized for his advocation for the inheritance of acquired characteristics,[12] made important contributions to evolutionary theory—the greatest being his recognition that the environment plays a critical role in evolution. Georges Cuvier (1769–1832), widely considered the founder of comparative anatomy and a leading proponent of catastrophism, had in his cabinets at the Muséum National d'Histoire Naturelle (National Museum of Natural History) in Paris the remains of so many animals from around the globe that he confirmed many fossilized bones coming to the museum were of animals no longer living. He was therefore the first person to document the existence of extinct species; as it was in the beginning, then, it no longer is. Of course, I wouldn't want to omit Alfred Russell Wallace (1823–1913), who independently came up with the concept of natural selection, spurring Darwin to finally write his book!

According to the historian of science Arnaud Hurel (2018), the early nineteenth century found scholars in two broad fields—the natural sciences and the human sciences—each with its own methodologies, working independently on the question of human antiquity. Hurel points out that, by historical convention, 1859 is the year these two fields became united in their view that humans had geological antiquity—and for two reasons. First, in 1859, both the Société Géologique de France (Geological Society of France) and the Société d'Anthropologie de Paris (Society of Anthropology of Paris, founded in 1859 by the physician and bioanthropologist Paul Broca [1824–1880]) published papers proving beyond all shadow of a doubt the existence of prehistoric humans. This was due to the discovery at multiple sites, excavated under controlled conditions, of stone tools and other clearly humanly made artifacts in association with extinct animals.

In contrast to the acceptance of these artifacts, however, human skeletal remains from prehistoric contexts were still controversial in the scientific community. This is not to say that they had not been found! In 1823, the theologian and geologist William Buckland (1784–1856) found a Paleolithic skeleton, missing its skull (but replete with red ochre-stained bones), in Paviland Cave, outside of Swansea, Wales. He erroneously determined that this skeleton of a young adult male was that of a woman of ill repute whom he associated with a nearby Roman encampment. He referred to this clearly scandalous person as the "Red Lady," due to the red ochre staining of "her" skeleton.[13] Other finds were made at Grande Grotte de Bize in France (1827), and Neandertals were recovered in Belgium and Gibraltar in 1829/1830 and 1848, respectively. However, the significance of these finds at first went unrecognized (chapter 3).

The second reason Hurel cites 1859 as a landmark year for human prehistory is due to the September meeting of the British Association for the Advancement of Science in Aberdeen. The highlight of this meeting was Charles Lyell announcing his acceptance of the existence of human prehistory. What I find fascinating is that all of these events occurred *before* the publication of Darwin's *On the Origin of Species by Means of Natural Selection*, which did not appear until November of that year!

Discoveries of prehistoric human remains from more controlled excavations soon followed. In 1861, the paleontologist Édouard Lartet (1801–1871)[14] documented that human remains, unfortunately lost before his arrival, had been associated with extinct fauna at the site of Aurignac, in the foothills of the Pyrenees. That same year he and his English collaborator Henry Christy (1810–1865) began to explore rock shelters in the Vézère Valley of the Périgord region in southwest France—the valley that is home to the site of Cro-Magnon and a multitude of other Paleolithic sites. In 1864, they found a human skeleton at La Madeleine just two miles from Cro-Magnon, but they hesitated to associate it with the Paleolithic, concerned that it could be a Neolithic ("New Stone Age") burial dug into Paleolithic levels. In 1866, human remains were recovered at Solutré in eastern France, about 65 kilometers (40 miles) north of Lyon. According to Morant, the poor stratigraphic controls of the early excavations at this site were such that one cannot discern the geological age or archaeological context of the human specimens;[15] this prevented their widespread acceptance as Paleolithic burials.

Finds like these were not greeted with enthusiasm. Many, like Solutré, were argued to be impossible to date or were thought to be Neolithic. Worse still, some were considered cases of fraud on the part of the excavator. A particularly notable case concerns the work of the customs officer and amateur archaeologist Jacques Boucher de Perthes (1788–1868) at the site of Moulin Quignon, near Abbeville in northern France. The political fallout resulting from an investigation into this discovery is discussed in detail by Trinkaus and Shipman, but a brief summary is warranted here.[16]

In 1863, Boucher de Perthes announced he had found a human mandible associated with stone tools at Moulin Quignon. The problem is that Boucher de Perthes's habit of paying workers for artifacts led to a cottage industry of at least some people in Abbeville making replicas of stone tools to pass off as the real thing. It is possible that this habit also led to their seeding the site of Moulin Quignon with a (decidedly fresh) human mandible. Today Boucher de Perthes is viewed as a hapless victim of fraud, rather than its perpetrator, but sadly he went to his grave believing the inquiry into the site's legitimacy ruined his reputation. The debate surrounding Moulin Quignon also took on an air of international partisanship, with British scholars tending to question its veracity and French scientists defending it. The whole affair ultimately faded in importance as more and more verified prehistoric human sites were found across Europe. It is into this complicated milieu that the site of Cro-Magnon comes to light in 1868, and there begins our tale.

I

Discovery

Lorsqu'on franchit pour la première fois, en chemin de fer, la distance qui sépare Limoges d'Agen, on ne peut se défendre d'un double sentiment de surprise et d'admiration en passant dans les défilés tortueux du Périgord noir, au fond desquels coule la Vézère. Les contrastes que présente cette vallée si fraîche avec les escarpements rocheux aux formes si bizarres qui la limitent brusquement des deux côtés, ménagent aux regards du voyageur le plus indifférent une succession d'effets aussi inattendus qui saisissants qui commandent son attention.

The first time one makes one's way by rail between Limoges and Agen, one cannot help but have a feeling of both surprise and admiration at the torturously narrow passageways of the Black Périgord, at the base of which runs the Vézère River. The contrasts this cool valley presents, with its strangely shaped rocky escarpments hemming one in so tightly on both sides, provides even the most indifferent traveler [with] a succession of visual effects so unexpected and striking that they command his attention.

—Louis Lartet, "Une sépulture des troglodytes du Périgord" (1868, translation mine)

Louis Lartet, like his father Édouard before him, spent years doing archaeological and geological surveys and excavating among the imposing limestone cliffs, cool rock shelters, and enchanting caves of the Vézère Valley of the Black Périgord in southwest France.[1] His description of the location remains as true today as it was when he wrote it in 1868. This valley, with one of the highest concentrations of Paleolithic sites in the world, is a lush paradise with steep cliff walls bedecked with broadleaf forests.

I first saw the valley in March 1994 when collecting data for my doctoral dissertation. My headquarters in the Périgord was its regional capital, the city of Périgueux, lying approximately halfway between Limoges and Agen. I was measuring and radiographing fossil hominins, and studying human fossils curated in Périgueux in what is now known as the Musée d'Art et d'Archéologie du Périgord (Museum of Art and Archaeology of the Périgord; at the time it was known as the Musée du Périgord). However, I also used Périgueux as my base of operations for day trips to museums in the nearby towns of Montauban, Brive-la-Gaillarde, and fatefully, Les Eyzies-de-Tayac-Sireuil (abbreviated to "Les Eyzies" from this point forward), home of the Musée National de Préhistoire (National Museum of Prehistory) and the Cro-Magnon rock shelter featured so prominently in this book (figure 1.1).

I still remember taking the train to Les Eyzies from Périgueux on that chilly, foggy March morning almost 30 years ago. Calling it a train is a misnomer because it was a small single-car tram, and for much of the morning's journey, I was its sole passenger. When only the driver (a compact, muscular man) and I remained, I struck up a conversation with him. I remember him as a man of few words, but friendly enough, and I tried not to show how shocked I was that he didn't know about the hominin fossils housed in the museum at Les Eyzies—a museum that he had surely driven past hundreds of times. As our little tram took its circuitous route toward and then over the Vézère, steep limestone cliffs rose on either side of us and the swiftly flowing river came into view. Early in the spring most of the trees were devoid of leaves, and I could see the craggy limestone behind them. Here and there I could make out the entrances of caves perhaps 30 meters (100 feet) above us. What struck me at the time is that people continue to live in the shelter of the cliff faces—many houses are built right into the rock face, with the cliff forming the back wall and often part of the roof of their homes. Despite this obvious sign of a human presence, the Vézère Valley has a wild, unspoiled feel to it, although I have been told by locals that most of its cliff faces were clear-cut of their trees in the early twentieth century.

When my tram stopped in the hamlet of Les Eyzies, I bade goodbye to the driver and stepped onto the platform, portable X-ray machine in tow. I started to make my way toward the museum—there is only one main road in town, so it is impossible to get lost. The first thing I noticed, to my left, was the Hôtel Cro-Magnon, located only 30 meters (100 feet) downriver from the main Cro-Magnon rock shelter itself. I'm sure all paleoanthropologists chuckle to

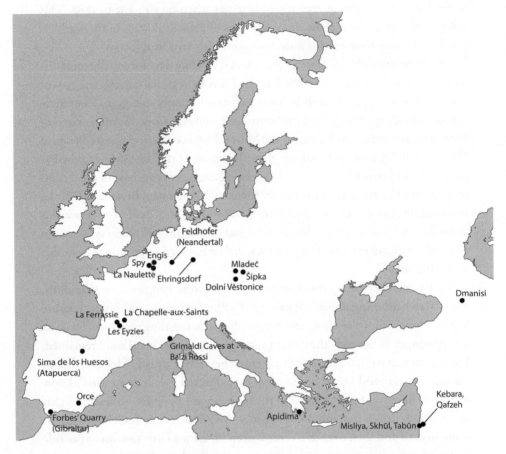

FIGURE 1.1 European and West Asian locations and sites mentioned in chapters 1 through 4. Sites in and around Les Eyzies include Cro-Magnon, Abri Pataud, Font-de-Gaume, Les Combarelles, Grotte de Bernifal, La Madeleine, La Rochette, Lascaux, Regourdou, La Balutie, and many others.

Source: Author's map.

themselves when seeing the bold white capital letters spelling out "HOTEL CRO-MAGNON" above the hotel's roof (figure 1.2), somewhat reminiscent of the Hollywood sign, but with a more paleontological flair. My family and I have stayed in the hotel, and it has French country charm and a welcoming feel. I love the hall on the second floor featuring the cliff face as its back wall—for like so many buildings in the Vézère Valley, it is built directly against the cliff.

FIGURE 1.2   Hôtel Cro-Magnon, July 2015; view from the train station in Les Eyzies.

Source: Photo by author.

The second thing I noticed as I walked through Les Eyzies was the imposing sculpture of a Neandertal high up on a ridge overlooking the village. This statue, *l'Homme primitif* (the primitive man) was created by the noted French sculptor Paul Dardé (1888–1963) in 1931. For 90 years, this imposing figure, with its slumped posture informed by the paleontologist Marcellin Boule's (1861–1942) view of Neandertals as maladapted evolutionary dead ends (chapter 3), has looked out over the Vézère Valley from high above the town in the courtyard of the Musée National de Préhistoire. I have been told that this statue was not popular with the locals and was treated with derision when first erected (much like the Eiffel Tower). It is now much beloved and is one of the most frequently photographed sights in town.

The final thing I will mention about my first trip to Les Eyzies is that I quite erroneously thought the town was dead. Almost all the shops were shuttered, and very few people were on the streets. I wondered if, as in many rural parts of the United States, the young people had gone elsewhere for jobs. Years later I discovered that I was there in the off-season. In the summer, the town is bursting with tourists (the local joke is that in summer there are more Dutch than French people in the Vézère Valley), and restaurants and bars have near-capacity crowds (at least this was true pre-COVID-19). Much of the attraction of vacationing in the region is nature (kayaking or canoeing on the Vézère, hiking along the craggy ridges to take in their panoramic views), but prehistory remains a huge draw. In addition to the Musée National de Préhistoire, Les Eyzies is also home to an interpretive center dedicated to prehistory and human evolution; it was there that I was humbled to learn that chimpanzees do better on some intelligence tests than humans! The town also boasts multiple prehistoric sites open to the public, including Cro-Magnon, Abri Pataud, Font-de-Gaume, Les Combarelles, and Grotte de Bernifal—the last three replete with Paleolithic parietal (wall) art. In Les Eyzies and its surrounding area there are no fewer than 37 rock shelter and cave sites with Upper Paleolithic art, and many more Paleolithic habitation sites in which no parietal art was found (there are ever-so-subtle traces of prehistoric art on the ceiling of the Cro-Magnon rock shelter itself). About 8 kilometers (5 miles) outside of Les Eyzies one finds the delightfully campy Préhisto Parc, with life-size dioramas of Paleolithic peoples engaged in a variety of behaviors, such as sneaking up on a woolly rhinoceros ("I'll take 'Things I would never do' for $1,200, Alex"), burying their dead, or creating art, all set in lush natural surroundings and guaranteed safer than Jurassic Park! Finally, about 24 kilometers (15 miles) upriver from Les Eyzies, on a hill just outside the town of Montignac, lies the world-renowned Paleolithic art site of Lascaux. Although the actual cave remains closed to the public due to damage to the paintings caused by visitors' exhalations, an actual size, nearly perfectly rendered 3D model of the cave complex and its art (Lascaux IV) has been open to the public since 2016. (I was honored to be given a private tour of this wonderful facility just a few months prior to its opening.) Two Paleolithic sites at which I have worked with principal investigators Bruno Maureille and Aurélien Royer—Regourdou (open to the public) and La Balutie (hard to access and closed to the public)—are also on the same hill as Lascaux. Prehistory clearly remains a touristic draw in the Vézère Valley.

DISCOVERY OF CRO-MAGNON REMAINS

In 1863, the railroad—that marvel of nineteenth-century travel and commerce—came to Les Eyzies, along the new Limoges-to-Agen line. According to Louis Lartet, to make room for the tracks in the narrow Vézère Valley just upriver from the Les Eyzies town center, some of the talus slope, a large limestone bloc, and part of a large overhang at the northeastern Cro-Magnon shelter were removed in 1863.[2] However, most of the Cro-Magnon shelter remained undisturbed, and the paleo-anthropologist Dominique Henry-Gambier says that in early 1868 it was still filled with (literally) tons of sediment—a massive deposit of dirt some 17 meters (56 feet) wide and 6–7 meters (20–23 feet) deep![3] Archaeologists *love* sediments! Archaeological sites tend to be sediment traps; if sediments are not trapped but washed away, the archaeological materials tend to be washed away as well.

For Cro-Magnon, the sediment trap would be forever transformed in March 1868, when a new train station for Les Eyzies and a road to serve it were built. Some 130 meters (430 feet) southeast of the new station, railway workers began to clear land for the road, and this required removing the sediment from the main Cro-Magnon shelter. In the course of clearing the land for the road, Lartet noted that the workers first removed detritus about 4 meters (13 feet) deep from Cro-Magnon.[4] They then began removing a rocky bank at the newly cleared area (some large blocks were removed with dynamite). Underneath this bank they ran into a motherlode of prehistoric debris. Most important, they uncovered evidence of human burials, including a nearly complete skull. These were no ordinary burials either. They were associated with stone tools, with shell beads that looked as if they had been strung together in necklaces or bracelets, and with broken animal bones, including those of reindeer (*Rangifer tarandus*). Later, in 1893 or 1897, a reindeer rib was discovered at Cro-Magnon onto which had been carved an anthropomorphic figure.

The importance of this find was not lost on two local entrepreneurs (Mr. Berton-Meyron and Mr. Delmarès) who had secured for the railroad company the property on which the station and road were to be built. In today's interglacial climate, reindeer are not found in France; instead they live in much colder regions. The finds at Cro-Magnon were therefore immediately recognized as having come from the last Ice Age. It had only been three years earlier that Louis Lartet's father, Édouard, announced the discovery at La Madeleine (just 3.2 kilometers [2 miles] upriver from Cro-Magnon), a mammoth tusk with an engraving of a mammoth on it—exciting proof of human coexistence with extinct fauna, a huge splash in the press. The people in the valley were therefore very much on the lookout for more such finds.

According to Lartet's description of the Cro-Magnon discovery, published in the *Bulletins de la Société d'Anthropologie de Paris*, we are fortunate that Berton-Meyron and Delmarès immediately asked Alain Laganne to come to the site.[5] Laganne was a local amateur geologist and archaeologist who had been called away to Bordeaux on business. Although not formally trained, he had much practical geological and archaeological experience and for years had been the foreman for Louis's father Edouard and his British colleague Henry Christy during their surveys of the Vézère Valley. Despite his lack of a university degree, Laganne was no slouch—Lartet mentions that by comparing erosion of limestone in locations where said erosion had occurred since a known date, Laganne had estimated that the limestone in the Vézère Valley erodes at a mean rate of about 15 millimeters (about half an inch) every twenty years. Although imperfect, this is an admirable attempt to quantify and extrapolate geological processes over time.

Upon receiving the news of the Cro-Magnon discovery, Laganne left Bordeaux and quickly returned to Les Eyzies. Some days later, he exhumed two more crania and several other fragmented human skeletal remains from what was left of the Cro-Magnon deposits, as well as more worked reindeer bone and flint tools. Finally, here at Cro-Magnon was the evidence the scientific world had been waiting for—human bodies associated with prehistoric tools and extinct (or locally extinct) animals such as reindeer, mammoth, cave lions, and cave bear.

This finding was, in a word, huge. Laganne contacted the historian Victor Duruy (1811–1894) in Paris, telling him that this fantastic discovery must be verified by formally trained experts. Duruy, who was the French education minister at the time, agreed; it is he who decided to send Louis Lartet to the site, apparently at the suggestion of Édouard Lartet.

When Lartet arrived at the site, much of the sediment from within the Cro-Magnon rock shelter had already been removed, either by the railway workers or by Laganne and his team. In Lartet's own account of the excavation, the first thing he noticed was that the roof of the shelter had a deep fissure in it, and was therefore at risk of collapse, so he installed a support column to ensure that the roof would not fall on him and his team as they worked. Having excavated rock shelters myself, I get this—I would rather not become part of the archaeological record by having a multiton shelter roof collapse on me. To secure the support pillar, Lartet needed to dig down to the bedrock, so he excavated four layers of burned material he interpreted as hearths. At the bottom of these layers was an elephant tusk (another indication of the site's prehistoric nature) that he and a priest (Father Sanna Solaro) removed. Once the support column was in place, his team carefully excavated the remaining layers.

Lartet noted that the bedrock, back wall, and roof of the Cro-Magnon shelter were composed of ancient limestone dating to the Cretaceous Period (the last age of the dinosaurs), based on the presence of fossilized coral and bryozoan colonies.[6] The bed over which the rock shelter lies included a brachiopod index fossil[7] that today is referred to the species *Cyclothyris vespertilio*, an extinct sessile, bottom-feeding sea creature.[8] Thus the rocky crags of the Vézère Valley are in fact an ancient seabed that was geologically lifted 100 meters (330 feet) above sea level. Over millions of years this originally marine limestone has slowly been

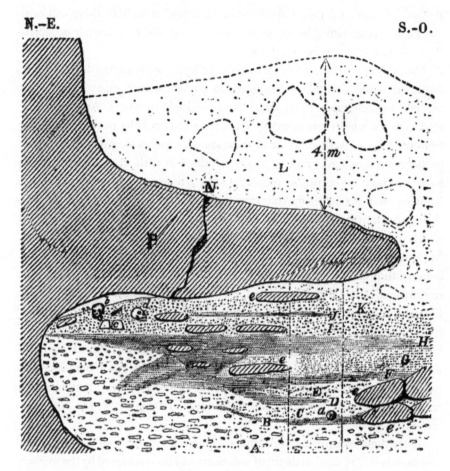

N.-E.   S.-O.

4 m

L

N

P

e

K

j

i

H

e

G

F

E

D

a

B   C

A

FIGURE 1.3 Reconstructed vertical profile of Cro-Magnon site, from northeast (L) to south-west (R). A = Basal (sterile) layer; B = First occupation layer; C = Sterile layer; D = Second occupation layer; E = Limestone layer reddened from heat; F = Third occupation layer; G = Reddish dirt layer; H = Thick layer including many occupations; I = Yellowish dirt layer that yielded the skeletons; J = Thin layer with hearths; K = Limestone scree; L = Removed talus slope; N = Fissure in shelter roof; P = Cretaceous limestone roof; a = Elephant tusk; b = "Old Man" skeleton; c = Stone block; d = Human bones; e = Blocs of roof fall.

Source: Louis Lartet, "Une sépulture des troglodytes du Périgord," *Bulletins de la Société d'Anthropologie de Paris* t. 3 (1868).

eroding to form caves and rock shelters primarily due to the natural acidity of rainwater and the action of ice.

Lartet's Cro-Magnon layers are depicted in figure 1.3. From the solid limestone bedrock at the floor of the shelter up to the top, he began with the basal, and therefore oldest, level. This layer is archaeologically sterile (no human remains or artifacts). Given its depth at about 70 centimeters (28 inches), these sediments had been accumulating in the shelter long before the first human occupants (Lartet's "reindeer hunters") arrived.

Above these basal sediments are a series of layers, some archaeological, others sterile (including roof falls)—but nearer to the shelter roof are a series of nearly continuous occupation layers that include multiple fireplaces. Artifacts recovered from these layers are primarily made from river cobbles people brought up from the Vézère riverbed. Many of these are what archaeologists call "scrapers," probably used for tasks such as scraping hides, but with sharp edges that may have been used for cutting or incising. Other artifacts in this layer are rounded hammerstones for either making stone tools or breaking open bones to access their fatty marrow. Still others are cores from which smaller flakes or blades were struck (the removed flakes or blades were then fashioned into various tools).[9] Most tools are made from local quartz and granite. Flint was also used, and outcrops of flint are a common occurrence in Périgord limestone, but it is a rarer raw material. If we use the nearby site of Abri Pataud (near the center of Les Eyzies) as a guide, some flints were probably local, but others were acquired from sources farther away. My late Tulane colleague (and lovable curmudgeon) Harvey Bricker (1940–2017) spent years excavating in the Abri Pataud. He found that some of the flint used was locally sourced in the Vézère Valley, but other tools were made on flint from the Couze Valley, 20–25 kilometers (12–15 miles) away, and still others were made on flint from north of Bergerac, 35–40 kilometers (22–25 miles) away.[10]

Above these occupation levels (indeed, above the shelter roof), an additional 4–6 meters (13–20 feet) in depth of sediments had been removed by the workers. Perhaps this material was what archaeologists refer to as "overburden"—sterile layers that overlie the archaeological material one wants to access—but it is also possible that evidence of later Paleolithic, Mesolithic, Neolithic, Bronze Age, or Iron Age deposits were simply discarded without study—we will never know. The silver lining for Lartet was that this massive quantity of sediment above the occupation levels should satisfy the scientific community that the archaeological layers lying beneath it were very ancient indeed.

The human skeletons, representing at least four adults and a child, were found at the back of the cave just beneath the shelter roof (figure 1.4). The skull of the most famous of the individuals, the so-called *vieillard*, or "old man" of Cro-Magnon (Cro-Magnon 1; figure 1.5), had been exposed on the surface at some point postburial, given the stalagmites adhered to his cranium. Lartet reported that the bones of the other individuals were all found within a 1.5-meter (5-foot) radius of the old man's skull. Just to the left of his skull, the skull of a

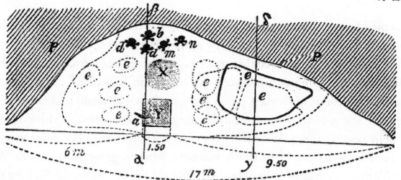

P. Calcaire crétacé.
X. Portion centrale et la plus épaisse de la couche H (voy. fig. 5).
Y. Base du pilier de soutenement.
a. **Défense d'éléphant.**
b. **Crâne de vieillard.**
d. **Ossements humains.**

e. **Dalles détachées de la voûte à différentes époques.**
m. **Squelette de femme.**
n. **Ossements d'un enfant.**
αβ. **Direction de la coupe de la fig. 5.**
δγ. **Direction de la coupe de la fig. 7.**

FIGURE 1.4 Reconstructed planar view of the site of Cro-Magnon, from northwest (L) to southeast (R). P = Cretaceous limestone; X = Thickest portion of layer H from figure 1.3; Y = Base of the support post; a = Elephant tusk; b = Skull of "Old Man"; d = Human bones; e = Blocs of roof fall; m = Cro-Magnon 2 skeleton; n = Infant bones; α-β = Plane of view for figure 1.3.

Source: Louis Lartet, "Une sépulture des troglodytes du Périgord," *Bulletins de la Société d'Anthropologie de Paris* t. 3 (1868).

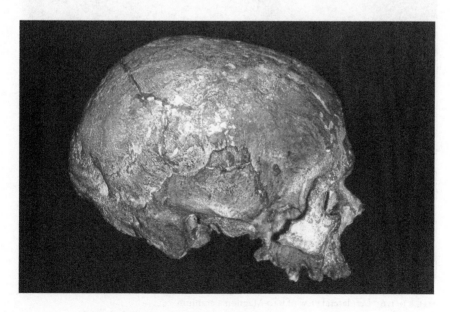

FIGURE 1.5 Right lateral view of Cro-Magnon 1 cranium.

Source: Courtesy of Erik Trinkaus.

presumed female (Cro-Magnon 2; figure 1.6) was recovered. Lartet noted that she had a wound, seemingly caused by a sharp implement, on the right side of her forehead. Broca described this injury as just above the external margin of her right orbit, measuring 33 millimeters (1.3 inches) in length and 12 millimeters (0.5 inches) wide at its thickest point. Doctors shown the skull told Lartet they suspected she had survived the wound for several weeks because there is evidence for bony healing along the edges of the injury. Right next to her skeleton were the remains of a neonate or nearly full-term fetus, begging the question: Was Cro-Magnon 2 pregnant when she died? Did she die in childbirth? Was she the victim of interpersonal violence, and if so, by whom? Broca also noted that a femur he attributed to the "old man" of Cro-Magnon shows evidence of a healed puncture in the shaft just above the knee.[11] He said this could be from a projectile weapon, or the horn of an animal such as an aurochs (ancestors of domestic

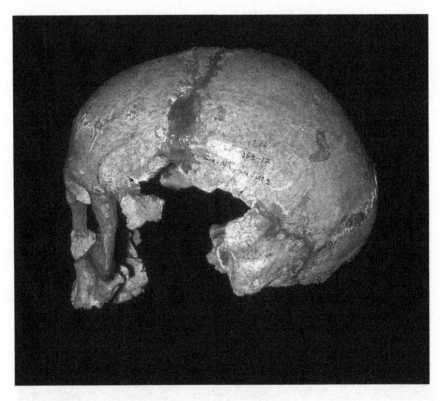

FIGURE 1.6 Left lateral view of Cro-Magnon 2 cranium.

Source: Courtesy of Erik Trinkaus.

cattle), or the tusk of a mammoth. In contrast, Dastugue rather convincingly argued that it was due to a systemic infection.[12] In either case, life appears to have been perilous for the residents of the Cro-Magnon rock shelter!

The other adult skeletons were presumed to have been males, and their remains are comingled (or were mixed during their discovery). In the midst of all these human remains were close to 300 pierced shells, the majority of which belong to the species *Littorina littorea*, a small marine snail. These were probably incorporated into bracelets, necklaces, or other decorative accessories associated with at least one (if not all) of the adult skeletons. The closest coast where one might find these marine animals was 150 kilometers (93 miles) away (and farther away still during their day because sea levels associated with the last Ice Age were lower), so either these reindeer hunters spent time on the coast or engaged in trade with those who did. Lartet also recovered an ivory amulet with two holes pierced in it not too far from where the human skeletons were recovered. Laganne had found a similar smaller amulet, and a schoolteacher (Mr. Grenier) gave yet another to Lartet that he claimed had been found by one of his students while visiting the site. Pierced animal teeth and worked reindeer antler were also recovered in the vicinity.

Given the depth at which the skeletons were found, the sediments and roof fall levels that had overlain them, as well as their association with extinct fauna and chipped stone tools, it made sense to Lartet that the human skeletons were from the Paleolithic, or "Old Stone Age." This makes them the first fossil *Homo sapiens* discovered in such an established archaeological context. I must point out, however, that scientists tend to be a skeptical bunch, and not everyone was convinced that the skeletons were that ancient. In a foreshadowing of events to come, in 1871 the prehistorian Émile Cartailhac (1845–1921) and the geologist Eugène Trutat (1840–1910) from the University of Toulouse published a paper in which they questioned the antiquity of the Aurignac human remains announced ten years earlier by Louis's father Édouard. Dominique Henry-Gambier could not help but notice that they published this paper only after the death of the influential Lartet *père*.[13] She also pointed out that Élie Massénat's 1872 discovery of a skeleton of Cro-Magnon age from the site of Laugerie-Basse, in the Vézère Valley just 2 kilometers (1.2 miles) upstream from Les Eyzies, provided Cartailhac with the opportunity to question not only its age but that of the Cro-Magnon skeletons as well.

Perhaps the most vocal critic of the antiquity of the Cro-Magnon skeletons was the archaeologist Louis Laurent Gabriel de Mortillet (1821–1898). At a Société d'Anthropologie de Paris meeting in April 1876, during discussion of the growing number of sites reporting Paleolithic human remains (and their use of shell beads), de Mortillet was particularly critical of Cro-Magnon. As transcribed by the conchologist Paul Fischer and published in the society's *Bulletins* in 1876, de Mortillet said that at first he too was caught up in the excitement of the Cro-Magnon finds and had no doubt of their antiquity. However, upon further investigation, he changed his mind. He noted that most of Cro-Magnon was excavated by railroad workers or Mr. Laganne, whom he referred to as a "simple researcher" (Ouch!). These unqualified excavators worked at Cro-Magnon for several days before the arrival of Louis Lartet. As a result, Lartet, whom he at least recognized as being a qualified expert, could only study the basal layers at the site (i.e., the ones below the human skeletons).

It was de Mortillet's opinion that the Cro-Magnon site had effectively been destroyed and could not be scientifically studied. He does, however, concede a later Upper Paleolithic (Magdalenian; discussed in chapter 2) presence at the site, but he posits that the human skeletons likely come from above the Magdalenian levels and may, in fact, be more recent (perhaps Neolithic, or "New Stone Age"). His *coup de grâce* is the following statement: "[Cro-Magnon] was evidently a burial, but we know of no certain burials from the Magdalenian period, while those from the time of polished stone [Neolithic] are abundant in caves."[14] Spoiler alert: we now have firm radiometric dates for the Cro-Magnon skeletons; they are much older than the Magdalenian!

Sadly, the site of Cro-Magnon has nothing left to tell us. In less than 50 years from the date of its initial discovery, the shelter was completely emptied of its contents. If you visit Cro-Magnon today, you will see a bare site dug down to its bedrock. The good news, however, is that within a few years additional modern human skeletons associated with extinct animals and stone tools came to light in Italy.

PALEONTOLOGISTS AND PRINCES:
DISCOVERIES AT THE BALZI ROSSI

"Mi scusi, parla inglese o francese?" ("Excuse me, do you speak English or French?") I ask at the ticket counter at the train station in Ventimiglia, a brightly painted Italian Riviera town clinging to the hillside at the mouth of the Roia River, and home to our next series of important Cro-Magnon sites. The gentleman behind the counter replies, "No." I think to myself that I could just about throw a rock into France from this station, and this guy doesn't speak French? Time to go with Plan B. "Si hablo español, puede Usted comprenderme?" (If I speak Spanish, can you understand me?) Again the response is monosyllabic, "No."

Perhaps you have guessed that once again the year is 1994. It is June, and I am at yet another location for my dissertation research. I have two work-related destinations while staying in Ventimiglia. The first is just outside of town: the Museo dei Balzi Rossi (Museum of the Balzi Rossi), home to several important Paleolithic skeletons from the Balzi Rossi (I'll explain what this is in just a minute). I also make day trips by train from Ventimiglia into France and on into Monaco, where more Balzi Rossi skeletons are housed at the Musée d'Anthropologie Préhistorique de Monaco (Museum of Prehistoric Anthropology of Monaco). Crossing all those international borders sounds more complicated than it actually is. The distance is only 20 kilometers (12 miles) and takes about 40 minutes by train.

But my business this evening is not research related. I'm planning ahead for the weekend; in two days I intend to take the train to Massa, a coastal city not far from Pisa, to visit the paleoanthropologist Vincenzo ("Enzo") Formicola. During my time in Europe, I have had a Eurail Pass, which has been a godsend these many months, but it has been recommended to me to reserve any seat I may need a couple of days in advance of departure. I have learned the hard way to make my reservations early.

Unfortunately, I don't speak Italian, but with my fluent (albeit very Louisiana) French and all the years of Spanish I took from fifth grade through college (I really should speak Spanish better than I do), if someone speaks slowly to me in Italian, I can usually get the gist of what they're saying. So despite his saying no, I show him my pass and proceed to ask him in Spanish if I can reserve a ticket to Massa for Friday evening. He replies that I do not need to reserve a ticket; I can come on Friday afternoon, and there will be room on the train. I ask could I please do it now while I am here at the station? This question is apparently more than he can take; I've struck a nerve. He launches into a scalding lecture in rapid-fire Italian. I don't quite catch everything he says, but I do get that now is the time in Italy when people go home to their families to eat dinner. I apologize. I don't want to be a culturally insensitive American; I am an anthropologist after all! I figure this guy is just about ready to end his shift to head home, so I decide to go get something to eat myself.

A couple of hours later I happen to pass by the station. Maybe someone new is at the counter who will be willing to help me. I walk in and the same guy is behind the counter. I say to him (in Spanish): "I thought you went home for dinner." He responds "no," that he has to work this evening. I ask him if I can reserve my ticket now? His response will probably not surprise you. "No."

The Grimaldi Caves are a cluster of caves and rock shelters on the Italian Riviera,[15] just 400 meters (a quarter of a mile) from the French border, on high cliffs overlooking a stretch of the Mediterranean coast more akin to Big Sur than to Miami Beach. The nearest towns are Menton, France, to the west, and Ventimiglia, Italy, to the east. The caves lie in a suburb of Ventimiglia known as Grimaldi di Ventimiglia, named for the powerful Grimaldi family. This family is associated with the former Republic of Genoa and the current Principality of Monaco, where by law the royal family must use the Grimaldi surname. Indeed, much of the funding for the excavation of the caves at the turn of the last century was provided by Prince Albert I of Monaco.

The geological feature in which the Grimaldi Caves are found is known as the Balzi Rossi ("Red Cliffs" in Italian), so named because the limestone has iron inclusions that give them a ruddy color. The caves are rich in archaeological deposits, and people have been excavating there since at least the early nineteenth century (figure 1.7). Given their proximity to France, and the fact that their initial study was conducted by francophone scholars (some funded by the Prince of Monaco), the caves are frequently discussed in the literature by their French names rather than their Italian ones. From west to east, the major Grimaldi Caves (with their Italian names first, followed by French and English translations, when appropriate) are Grotta dei Fanciulli (Grotte des Enfants or Cave of Children),

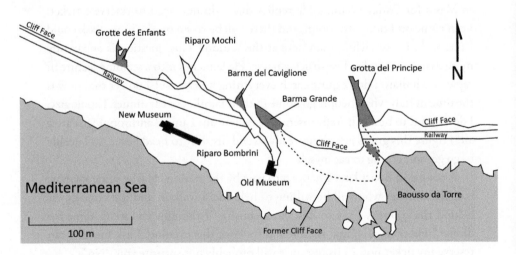

FIGURE 1.7   Map showing the locations of the major caves and rock shelters at the Balzi Rossi.

Source: Author's map.

Riparo Mochi (l'Abri Mochi), Barma del Caviglione (Grotte du Cavillon), Riparo Bombrini, Barma Grande (Big Cave), Barma di Baousso da Torre (Baousso da Torre is "Knob [small hill] of the Tower" in Occitan; its more standard Italian spelling is Bausu da Ture), and Grotta del Principe (Grotte du Prince or Cave of the Prince).

By 1868, the year the Cro-Magnon skeletons were discovered, locals and foreigners alike had been excavating in the Grimaldi Caves for decades. Prince Florestano I of Monaco recovered animal bones and stone tools there in 1846 (giving the Cave of the Prince its name), and Paul Broca dug in the Grimaldi Caves in 1865. None of these early excavations was terribly systematic, and few finds were reported in the scientific literature. This changed with the work of a young French physician named Émile Rivière (1835–1932).

In January 1870, Rivière moved to the French Riviera town of Menton, just across the border from the Balzi Rossi, to open his medical practice. By October 1870, however, he was spending much of his free time excavating in the Grimaldi Caves. He was soon rewarded with some magnificent finds, beginning in March 1872 with the primary burial of an adult human skeleton from the Barma del Caviglione. This individual was recovered a full 6.55 meters (21.5 feet) below what was then the cave floor. Its entire body was stained with red ochre, and it wore a cap, or some form of headdress, adorned with pierced shells and deer teeth. The occupation level directly overlying the Barma del Caviglione skeleton had hearths with worked stone tools and the bones of extinct animals (cave bears, cave lions, and cave hyenas), so Rivière was confident that this, like the then-recent finds at Cro-Magnon, was Paleolithic in age. The skeleton would become known in France as *l'Homme de Menton* (Menton Man). The entire specimen was removed *en bloc* and has been in Paris since 1872, where it long awaited formal scientific analysis, which was finally published in 2016.[16] Perhaps the most interesting conclusion drawn in that book is that the bioanthropologist Jaroslav Brůžek, using new techniques to analyze the specimen's pelvis, has determined that Menton Man was female!

More finds of Paleolithic human skeletons soon followed. In 1873 Rivière discovered skeletons of two adults (BT1 and BT2) and one adolescent (BT3) while excavating the Baousso da Torre. The adolescent was buried face down, which is unusual (we'll see this again at a different site in chapter 9). The two adults were covered in red ochre. For more than 80 years, BT1 and BT3 were thought to have been lost to science, but they were rediscovered in 2008 in the collections of the Musée Lorrain in the French city of Nancy by Sébastien Villotte and Dominique Henry-Gambier.

Rivière next discovered two child burials at the Grotte des Enfants (giving the site its name) in 1874 and 1875. These two skeletons (Grotte des Enfants 1 and 2), the first four to six years of age and the second about four years old, were found buried side-by-side about 2.7 meters (9 feet) below what was then the cave floor. Each had been wearing clothing adorned with pierced shells. Rivière argued that the nature of the treatment of the dead was different for adults and for children. Children were intentionally buried in a pit, he said, whereas adults were covered with ochre and left on the cave floor.

In 1883, Rivière began excavating at the Barma Grande and recovered archaeological material. The most important find, however, was made by the foreman of quarry workers employed at the site, Mr. Louis Julien, who was known to trade in antiquities—a huge "red flag" for archaeologists. In February 1884, Julien discovered a beautifully preserved large male skeleton (Barma Grande 1) some 8.4 meters (28 feet) below the cave surface. He was lying on his back—with three big flint flakes, one at the top of his head and the other two placed on the shoulders as if they were epaulets, and a thick "cap" of red ochre laid on his head—indicating to Rivière that he had been intentionally buried by members of his own group or family. Julien made arrangements to have the skeleton sent to a small museum in Menton. Unfortunately, only part of the material was sent; the rest was destroyed in a dispute over who had rights to the skeleton.[17]

Despite the depth at which these specimens were found, their geological age was controversial. One problem is that at many, if not all, of the Grimaldi Caves there was mixing of archaeological levels via postdepositional processes. Archaeologists today refer to the study of these and other processes that impact site integrity as taphonomy. Even though he supported an ancient age for the caves, Rivière acknowledged this phenomenon. In dismissing de Mortillet's argument that the burials were Neolithic or Magdalenian, Rivière argued that the entire sequence in the Grimaldi Caves, top to bottom, was Pleistocene in age. Yet, at the same time, he admitted that tools from the Mousterian (associated with Neandertals in Europe; chapters 2 and 3) and the much later Magdalenian were often found side-by-side in the same levels with no evidence of any disconformity between them! Despite this mixing of levels, most scientists in the 1870s were convinced the Grimaldi skeletons were Paleolithic; only a handful of scholars (most notably Moggridge and de Mortillet) thought otherwise.

In 1892, an unfortunate turn of events occurred at the Barma Grande. The landowner, Francesco Abbo, had destroyed Baousso da Torre by quarrying limestone in it, and he began to quarry in a haphazard way in the Barma Grande as well. On February 7, 1892, his workers uncovered a second large male skeleton (Barma Grande 2, BG2) near the cave entrance, not far from where Julien had

FIGURE 1.8   The Barma Grande 3 and 4 adolescent skeletons in situ.

Source: Courtesy of Fabio Negrino and the Istituto Internazionale di Studi Liguri.

found the first skeleton eight years earlier. A few days later, two additional skel-etons (BG3 and BG4) were discovered next to it in what appeared to be the same burial pit (figure 1.8). News of the find quickly spread, and the biological anthropologist René Verneau (1852–1938) was tasked by the Muséum d'Histoire Naturelle de Paris to study them. When he arrived, he noted that all three indi-viduals had been laid in a single burial pit, although most of the pit walls had been removed before his arrival.[18]

The British archaeologist Sir Arthur Evans (1851–1941) was also at Barma Grande around the time of discovery and published the following observations in 1893. First, BG2 was an impressively large individual. According to Evans, the length from his heel to his shoulder was 1.85 meters, almost 6 ft. 1 in., with-out taking into account his head or neck![19] A long flint knife lay close to his left hand, and about his head and neck were ornaments of pierced shell, teeth, and bones. The skeleton next to him (BG3) was recognized as a late adolescent female. She was not as decorated as BG2, but she was also holding a flint blade. The third skeleton (BG4), behind the presumed female, was another adolescent.

This individual was said to have had a flint blade located near the head. Verneau suggests that BG4 was wearing a necklace made of pierced and incised red deer canines and shells.[20] Verneau had sexed the two adolescents as female (BG3) and male (BG4), but his assignments are uncertain because subadult skeletons have yet to develop their secondary sexual characteristics. Unfortunately, we can no longer assess their sex using their pelves; these were destroyed in World War II. That said, Formicola and Holt report that genetic analyses confirm that the large Barma Grande 2 skeleton is male, and preliminary results suggest that both adolescent skeletons are female.[21]

Evans relates an unfortunate story about his second visit to Barma Grande sometime later that month. He says Mr. Abbo had assured him he was safeguarding the Barma Grande ornamental pieces in his house, but when Evans arrived there, they and several other artifacts had disappeared. To make matters worse, Evans claims that although Abbo had been making quite a bit of money charging visitors a franc a piece to visit the site, he "did practically nothing to protect the skeletons, which in a few weeks' time were so trodden under foot as to be almost past recognition."[22] Verneau is less harsh with the landowner, stating that in spite of Abbo's attempts to set up a barrier at the cave entrance many curious visitors showed up at the site, causing damage to the skeletons. (Verneau makes no mention of an entrance fee.)[23] Throughout this time, quarrying at Barma Grande continued unabated, and in January 1894, Abbo recovered a fifth (BG5) skeleton near the back of the cave. Shortly thereafter a sixth skeleton (BG6) was recovered, one Verneau interpreted as burnt, a conclusion disputed by more recent scholarship.[24]

Good news finally emerged around this time as well. Prince Albert I of Monaco, the grandson of Florestano I, had, like his grandfather, excavated in the caves bearing his family name back in 1882. Increasingly worried about the destruction of the caves, in 1895 he funded scientific investigation of the sites, especially its human skeletons. He put a priest (and Monaco's chief librarian), Léonce de Villeneuve (1858–1946), in charge of the project. This may seem an odd choice, but de Villeneuve had extensive archaeological experience. To help with the work, de Villeneuve brought onboard many of the top scholars of the day, all of whom happen to be French: Émile Cartailhac, an archaeologist from the University of Toulouse (discussed earlier), charged with the task of exhuming any human skeletons; Marcellin Boule (1861–1942), a paleontologist at the Muséum National d'Histoire Naturelle in Paris and a former student of Cartailhac, whose primary responsibility was to examine the fauna (he later became famous for his description of the important La Chapelle-aux-Saints 1 Neandertal); and the biological anthropologist René Verneau, who had already studied the previously excavated Grimaldi skeletons. This project, funded by the prince and led by the priest, continued until 1905, culminating in a series of ahead-of-their-time monographs published in 1906.

Beginning in 1900, this "Dream Team" hit the motherlode in the Grotte des Enfants. Cartailhac found conclusive proof that the inhumations are just that—intentional burials of prehistoric humans; but perhaps more important, Cartailhac, once skeptical of Paleolithic burial, was now convinced that the artifacts associated with these burials, and the layers from which they were drawn, were distinctly Paleolithic in character. Boule found that the relevant fauna from all levels of the site were Pleistocene, or "Ice Age" in character. This points to a group of Ice Age hunter-gatherers living in complex societies—exciting new examples of Cro-Magnon people.

There are new burials too—four, to be exact—from the Grotte des Enfants. They come from a host of different levels—one (Grotte des Enfants 3; now known to be later than the others) is only 1.9 meters (6 feet) below the surface. An adult female, she was buried with a wild boar jaw, several flint flakes, and two pierced shells. A second skeleton (Grotte des Enfants 4) was found in much deeper deposits, about 7 meters (23 feet) below the surface. A large adult male, he was not lavishly adorned, but pierced red deer teeth were recovered near his head, and there are pierced shells among his left ribs. Finally, two skeletons (Grotte des Enfants 5 and 6) were found deeper still, some 7.75 meters (25 feet) below the surface. Grotte des Enfants 5 is an older adult female; Grotte des Enfants 6 is an adolescent male. It appears that the female is an intrusive primary burial into the earlier burial of the adolescent male (it is uncertain how many intervening years there were between the two burials). He has four rows of pierced shells on his head, as if they were attached to some form of headdress, and several flint blades were found between the two skeletons. She has a pierced shell bracelet around her wrist, and pierced shells were found under her tibia.

The fauna Boule catalogs in this deepest layer include cave bears, cave hyenas, cave lions, rhinoceros, elephant, and hippopotamuses. This not only indicates Pleistocene age but suggests an earlier, warmer period within the Pleistocene than the so-called Reindeer Age.[25] To Boule this means the layer is very ancient indeed. The Grotte des Enfants 4 skeleton is associated with many of the same mammals. Boule had spent years excavating Mousterian sites in the Périgord, so he interpreted this to mean that these modern-looking skeletons may have been contemporaries of the Neandertals (or at least the earliest of the Grimaldi skeletons were only slightly later in time than their much more primitive-looking cousins).

Grotte des Enfants 3 is associated with red deer, fallow deer, and ibex—fauna that still inhabit the area today—but the big cats recovered in the deeper levels are absent. Although reindeer are also absent, the faunal profile of this level is similar to those in neighboring caves in which reindeer were recovered. As such, Boule characterized this most recent burial as from the Reindeer Age, roughly contemporary with the fossil humans from Cro-Magnon.

All the skeletons, including the three found in early, warmer-climate contexts, had been placed in distinct burial pits dug for them by their fellow group members. Verneau argued that it is no longer tenable to claim that Paleolithic hunter-gatherers did not bury their dead. He also critiqued Rivière, who had said the first skeleton at Baousso da Torre and the one from Barma del Caviglione had been left exposed on the cave's surface. Verneau pointed out that there has been nearly constant occupation of the Grimaldi Cave sites, with stacks of archaeological layers, and features such as fireplaces cutting into other, earlier occupation layers. Living next to a decomposing corpse is not something people tend to do, so he suspected that, similar to the burial pits he documented, the adult skeletons Rivière had recovered were also buried in pits.

Thus, by the early years of the twentieth century, the Grimaldi Caves had yielded multiple intentional burials by Paleolithic hunter-gatherers. These burials were recovered under controlled excavations by recognized experts. Given the ornamentation associated with them, despite their antiquity these ancient people were clearly intelligent beings who lived in complex and artistic societies. The next chapter explains how Paleolithic archaeologists study these ancient people.

2
Archaeology of the Ancients

It's June 18, 1999, about an hour after sunrise. This time of year the sun rises early, so I'm not yet fully awake. All I hear is the barely audible purr of the small automobile engine as we motor down a rural backroad. What's overwhelming is the morning light. It's almost blinding as it shines down on the smooth grass-covered rise some 50 meters (55 yards) to our left. To our right is the massive Pálava Hill (elevation 484 meters [1,587 feet]), the westernmost Carpathian mountain that dominates the landscape for miles. Aside from Pálava, the land is a rolling plain as far as the eye can see. We are in the greater Danube Valley.

Atop the rise to our left, a man and a woman are screening—buckets of dirt have been emptied into a large box with a screen bottom, and the two are now vigorously shaking the box screen. As they shake, dirt falls through the screen, leaving artifacts and a cloud of dust behind, so these two are obviously archaeologists. It reminds me of the scene in *Raiders of the Lost Ark* in which the archaeological crew is screening up on an Egyptian hill with the setting sun behind them.

I find it funny that we're watching people do archaeology because there are three archaeologists in the car with me. The Czech archaeologist Jiří Svoboda, an expert in lithic technology, is driving. We three passengers are all Americans: me, Olga Soffer, and James "Jim" Adovasio. Olga specializes in the Upper Paleolithic of Central and Eastern Europe; Jim is best known for having accidentally discovered an amazing pre-Clovis[1] site in western Pennsylvania called Meadowcroft Rockshelter. We have all been staying at the Paleolithic and Paleoethnology Research Center in the tiny Czech village of Dolní Věstonice, just 11.3 kilometers

(7 miles) north of the Austrian border. Jiří lives in Dolní Věstonice most of the year. I've been here for a few weeks studying the ribs and vertebrae of the Upper Paleolithic skeletons from Dolní Věstonice and the neighboring village of Pavlov. Olga and Jim are examining 31,000-year-old archaeological evidence for basketry and textiles from Dolní Věstonice—exciting stuff, and I've really enjoyed getting to know them.

This morning we're headed to Vienna to see a talk on Neandertals by the Croatian paleoanthropologist Jakov Radovčić. I've been looking forward to this trip all week.

It's Jiří who breaks the silence. "That's my friend's site up there."

"What kind of site is it?" Olga asks.

"It's a Roman fortification."

She exhales audibly in disgust as she says, "Current events!"

Paleolithic archaeologists study time periods that are so old that they really do make "ancient Rome" look like "current events." I'll demonstrate this with a simple analogy: our human lineage split from the chimpanzee-bonobo lineage around 8 million years ago.[2] Imagine that these 8 million years are instead the distance of the cross-country road trip from the Empire State Building in New York City to the Hollywood Walk of Fame in California—a road trip of some 2,785 miles. We leave New York City 8 million years ago as the first hominins. The earliest stone tools, known as Lomekwian, are about 3.3 million years old. Along our cross-country road trip we don't encounter these tools until we're in McClean, Texas, 74 miles east of Amarillo and just 35 miles west of the Texas-Oklahoma border. The earliest Oldowan tools, smaller than those of the Lomekwian, show up around 2.6 million years ago. We first encounter them in Santa Rosa, New Mexico, about 120 miles east of Albuquerque. At this point we're already two-thirds of the way through our journey.

In contrast, the maximum extent of the Roman Empire occurred under Emperor Trajan in the year 117 CE. We encounter this about 7/10 of a mile away from the Walk of Fame, as we turn from the exit off of the Hollywood Freeway onto Hollywood Boulevard itself! In this light, there's nothing "ancient" about ancient Rome. But while I laughed at Olga's joke, in the grand scheme of things, the sites at Dolní Věstonice are not so terribly ancient either. At 31,000 years old, we ran into them just 10.8 miles from our destination, as we passed the Monterey Park Golf Club in Los Angeles, 4.5 miles east of downtown.

THE HISTORY OF HUMAN TECHNOLOGY

In 1817, the Danish antiquarian Christian Jürgensen Thomsen (1788–1865) arranged artifacts in his museum according to his view of the history of human technology as a three-age system—a system still used today, albeit with modification. The earliest tools were made of stone; these he called Stone Age. Around 5,000 years ago, however, humans learn how to alloy copper with tin to make bronze; these belong to the Bronze Age.[3] Sometime later, around 3,200 years ago, humans created fires hot enough to smelt iron; thus begins the Iron Age.

In 1865 the English Baron John Lubbock (1834–1913) split the Stone Age into early and late components. The early Stone Age is marked by chipped stone tools; Lubbock called this the Paleolithic or Old Stone Age. Later Stone Age tools, made of smoothed or polished stone and often found with pottery, Lubbock named the Neolithic or New Stone Age. These distinctions remain with us, with the addition of the Mesolithic, spanning the phase between the Paleolithic and Neolithic, characterized by small, sharp pieces of stone called microliths.

There are a few caveats here. One is that in the Paleolithic we tend to focus on lithic (stone) artifacts simply because of stone's durability. However, our closest living relatives (chimpanzees and bonobos) make and use tools in the wild, primarily from perishable materials such as sticks and leaves; Occam's razor would suggest that early hominins also made tools from perishable materials—tools unlikely to appear in the archaeological record. These tools could very well date back to the last common ancestor of humans and chimpanzees more than 8 million years ago. Keep in mind, too, that some groups of wild chimpanzees use hammerstones and anvils (at times anvils made of stone) to break open nuts. Unfortunately, in the archaeological record, tools like this would be hard to distinguish from randomly trampled or bashed rocks. Unless the use of multiple hammerstones and anvils had been concentrated in the same spot for long periods, the probability of finding any site containing them would be vanishingly small indeed.

Human lithic technology that *can* be detected in the archaeological record has its humble beginnings about 3.3 million years ago at the site of Lomekwi, just west of Lake Turkana in Kenya (figure 2.1). Here some hominins figured out that sharp edges could be created by striking rocks together, and these rocks could then be used for a variety of tasks. By archaeological convention, the rock that is struck by a hammerstone is called a "core" and the piece that is knocked off is a "flake." Archaeologists call this process of chipping stone "knapping." There are

FIGURE 2.1 African and West Asian archaeological and paleontological sites mentioned in the book.
_____
Source: Author's map.

hammerstones and apparent anvils present among the artifacts at Lomekwi, and these appear to be much larger and heavier than those used by wild chimpanzees.[4] It is this Lomekwian industry that marks the beginning of the Paleolithic, which in terms of time comprises over 99 percent of the archaeological record.[5] The Paleolithic is said to have ended 11,700 years ago, with the beginning of the Holocene epoch, but in one sense it never really goes away. In the ethnographic present, humans are known to make expedient stone tools whenever they need a quickly and easily made sharp edge.

Once chipped stone tool use had been established, you might think it would be such a fundamental technological innovation that it would never disappear. Although we must be mindful of the adage that "absence of evidence is not evidence of absence," there seems to be about a 700,000-year hiatus in hominin stone tool use after the Lomekwian. Despite having multiple locales in eastern Africa with sediments dating to less than 3.3 million and more than 2.6 million years ago, we have no evidence for stone tools anywhere in the region for this period (table 2.1).

Around 2.6 million years ago, at sites like Gona in Ethiopia, we begin to find evidence of a new chipped stone tool industry—the Oldowan. Although based on the same principle of creating sharp edges by knocking rocks together, Oldowan tools tend to be much smaller, easily fit in the palm of the hand, and their makers appear to have had a better understanding of the conchoidal fracture mechanics of stone. It also seems likely that the makers of the Oldowan employed a freehand knapping technique in which the core is held in one hand and struck with a hammerstone held in the other. In contrast, the makers of the Lomekwian (species unknown) tended to employ either a passive-hammer technique, in which the core is held in both hands and struck against a large, hard, stationary object, or a bipolar technique, in which the core is placed on an anvil and struck with a hammerstone to knock off a flake. These two types of manufacture have much more in common with the way chimpanzees use stone tools in the wild, and the Lomekwian may therefore be a bridge between a nut-cracking technology with one more focused on creating sharp edges.

The Oldowan was first described by the archaeologist Mary Leakey (1913–1996), who discovered what were then the oldest-known stone tools in 1.8-million-year-old levels at Olduvai Gorge in Tanzania in the 1950s and 1960s. Leakey grouped the cores she found into a variety of tool types such as spheroids, choppers, scrapers, and discoids, based on shape and presumed function. Most Oldowan tools were unifacially worked (i.e., had flakes removed from only one side of the core), but a sizable proportion were bifacially worked (i.e., had flakes removed from two sides). We do not know that these tools were used in the way Leakey's names suggest, and many appear to have been multipurpose tools. Use-wear analysis tells us that in many cases the flakes were used as tools as well.

The earliest artifacts in Europe are Oldowan, including the 1.75-million-year-old Dmanisi site in Georgia[6] and the 1.4-million-year-old Orce sites in Spain.

TABLE 2.1 Chronology of Lower Paleolithic industries in the Old World (dates approximate)

Years BP	Africa	West Asia	Europe	East Asia
	to ca. 170 ka	*to ca. 300 ka*	*to ca. 140 ka*	*to ca. 55 ka*
600 ka	↑	↑	↑	↑
800 ka			Acheulean	
1.0 Ma				Acheulean (some places)
1.2 Ma				
1.4 Ma		Acheulean		
1.6 Ma	Acheulean			
1.8 Ma		Oldowan-like	Oldowan-like	
2.0 Ma				
2.2 Ma				Oldowan-like
2.4 Ma				
2.6 Ma	Oldowan			
2.8 Ma				
3.0 Ma	No stone tools			
3.2 Ma	↓			
3.4 Ma	Lomekwian	No stone tools	No stone tools	No stone tools

Note: The Middle Paleolithic was well established across the Old World by ca. 200 ka (200,000 years ago), but the Acheulean survived later than this in many locations. Ma = millions of years ago.

FIGURE 2.2 Acheulean hand ax of unknown provenience. It measures 10.8 centimeters (4.25 inches) in length, with a maximum width of 6.1 centimeters (2.4 inches).

Source: Photo by author.

However, another more technologically sophisticated industry, the Acheulean, first appears in Africa more than 1.7 million years ago, and it eventually spreads across Africa and into much of the rest of the Old World. These tools are almost all bifacially worked, and the hallmark of this industry is the hand ax, a teardrop-shaped biface that was likely the "Swiss army knife" of the Paleolithic—a source of smaller sharp flakes when needed, and a larger, more robust, cutting edge if the job called for one (figure 2.2). As their name suggests, these tools do not appear to have been hafted onto handles/shafts but were hand-held. It is also during this industry, about 700,000 years ago, that hominins discovered softer materials such as antler, bone, or wood could be used as hammers to knock smaller, thinner flakes off a stone core. This "soft hammer" technique persists throughout the Paleolithic.

Why is the hand ax industry called the Acheulean? The first hand axes known to science were discovered in France by Boucher de Perthes (chapter 1), but the physician Marcel-Jérôme Rigollot (1786–1854) later found many in the town of Saint-Acheul. The industry thereafter became known as the Acheulean (sometimes spelled Acheulian). It is replaced by more sophisticated technology around 200,000 years ago, although it survived later than this in many places (see table 2.1).

The Paleolithic, or Old Stone Age, is now further subdivided into three main phases: the Lower, Middle, and Upper Paleolithic. The Lomekwian, Oldowan, and Acheulean industries represent the Lower Paleolithic. These are industries in which most of the identifiable tool types are modified cores, with the proportion of bifacially worked tools increasing from the Lomekwian to the Oldowan to the so-called Developed Oldowan to the Acheulean. The technological innovation that separates the Middle Paleolithic from these Lower Paleolithic industries is that in the Middle Paleolithic most tools are made not on cores but on flakes.

As early as 500,000 years ago, knapping methods known as "prepared-core" techniques appeared, and by 200,000 years ago they were widely distributed across the western Old World. Adoption of these techniques marked the beginning of the Middle Paleolithic. The most famous of these techniques is the Levallois technique, named for the Paris suburb in which it was first experimentally recreated by modern-day knappers. In this process, an area of a core is carefully prepared using a soft hammer to knock off a series of small flakes in a radial fashion, turning the core in the hand after each strike such that the worked side of the core looks a bit like a tortoise shell. Once the area has been shaped to the knapper's satisfaction, the core is struck with a large hard hammer, and the preshaped side is dislodged from the core. Unlike in the Lower Paleolithic, this flake, not the original core, is now the desired tool. However, more pre-prepared flakes may later be dislodged from the same core, depending on its size.

There is variation both in space and time in Middle Paleolithic industries. In sub-Saharan Africa, for example, the most common Middle Paleolithic industry is called the Middle Stone Age (MSA). It has high levels of technological variability and shows similarities to the European Upper Paleolithic (chapter 7). In northwestern Africa, primarily in Morocco, Algeria, and Tunisia, and sometimes as far east as the western desert of Egypt, is a Middle Paleolithic industry known as the Aterian (figure 2.3), in which almost every tool has a "tang" on it—a small projection used to help haft the tool into a shaft. Unlike the Lower Paleolithic, in which most tools appear to have been handheld, in the Middle Paleolithic there is evidence for frequent hafting of stone tools into wood, bone, or antler shafts.

The Middle Paleolithic industry most relevant to the subject matter of this book is the Mousterian (figure 2.4), named for the Vézère Valley site of Le Moustier, first dug by Édouard Lartet and Henry Christy in the 1860s. In Europe, the Mousterian thus far appears to be solely the work of Neandertals. However, in

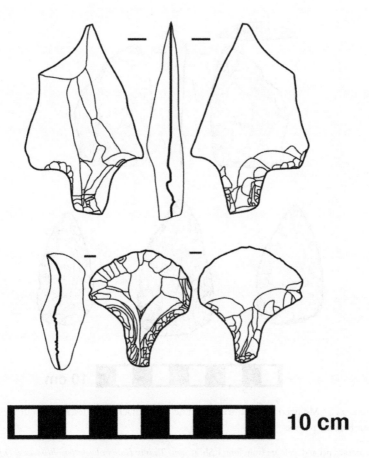

10 cm

FIGURE 2.3 Two Aterian lithic artifacts from Oued Djebbana, Algeria, each in three views. The top is a point; the bottom is a scraper.

Source: From John J. Shea, *Stone Tools in Human Evolution: Behavioral Differences Among Technological Primates* (Cambridge: Cambridge University Press, 2017), courtesy of John J. Shea.

the southern Levant, the Mousterian is associated not only with Neandertals but also with early modern humans at the Israeli sites of Qafzeh and Skhūl (and perhaps at Misliya Cave; chapter 4). In Europe, however, Cro-Magnons appear to have made only Upper Paleolithic tools.

The European Upper Paleolithic, the industry produced by Cro-Magnons, is characterized by long, parallel-sided blades that are "punched" off cores, as well as frequent use of bone and antler as raw materials. It is further subdivided into temporospatial industries. For the interested reader,

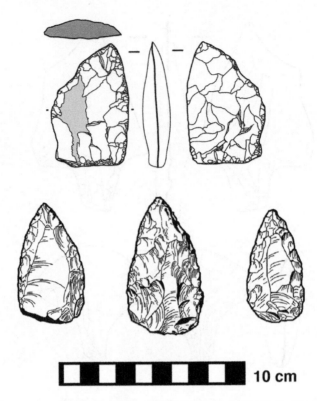

FIGURE 2.4 Mousterian lithic artifacts. The top artifact is a sidescraper (four views); the bottom three are points.

Source: Sidescraper and scale are from John J. Shea, *Stone Tools in Human Evolution: Behavioral Differences Among Technological Primates* (Cambridge: Cambridge University Press, 2017), courtesy of John J. Shea. The points are redrawn from M. C. Burkitt, *The Old Stone Age: A Study of Palaeolithic Times* (New York: New York University Press, 1956) by the author.

Box 2.1 discusses these Upper Paleolithic industries in more detail, and table 2.2 outlines regional/temporal patterning in European Upper Paleolithic industries up through the Neolithic.

DATING OF PALEOLITHIC SITES

How do archaeologists determine the age of a layer, artifact, or skeleton? There are two types of chronology: relative and absolute. Relative chronology is the statement that artifact or layer "A" is older than artifact or layer "B." This is done by

BOX 2.1. EUROPEAN UPPER PALEOLITHIC INDUSTRIES

The earliest Upper Paleolithic industries in Europe are grouped under the umbrella term **Initial Upper Paleolithic** (IUP) and show both Middle and Upper Paleolithic elements. Which hominin species are responsible for the IUP remains controversial. At Mandrin, France (**Neronian industry**), and Bacho Kiro, Bulgaria (**Bachokirian industry**), we now know the IUP makers (54,000–46,000 BP) were modern humans (table 2.2; chapter 8). A particularly controversial IUP industry is the **Châtelperronian**, a regional industry from France and northeastern Spain dating from about 44,000 to 40,000 years ago. Long considered an Upper Paleolithic industry because of its high proportion of blade tools, its type fossil, the Châtelperron knife, is a backed blade with a dulled side, either for the hand or for hafting to a shaft (chapter 10). It is also characterized by a significant number of tools made of bone and antler, as well as personal ornaments. There is debate as to whether the Châtelperronian was made by Neandertals, whether the Neandertals invented it, or if it represents imitation of modern human technology by Neandertals (chapter 8). Other IUP industries include the **Uluzzian** in Italy and Greece, the **Bohunician** in Czechia, and the **Szeletian** in Czechia, Slovakia, Hungary, and Poland (table 2.2). Here, too, there is no agreement on who is responsible for these industries—early modern humans or Neandertals (chapter 8).

Another Early Upper Paleolithic industry appears almost simultaneously across the breadth of Europe: the **Aurignacian**, dating from about 43,000 to 29,000 years ago. The type fossil of this industry is the split-based bone point, split at its base to facilitate hafting onto a shaft; it is almost certainly a projectile weapon (chapter 10). At many sites, the Aurignacian proper is preceded by an industry known as the **Protoaurignacian**, which lacks split-based bone points, instead having tiny curved blades called Dufour bladelets (chapter 8). The Protoaurignacian and the Aurignacian are thus far only associated with modern humans.

The **Gravettian** temporally overlaps with the Aurignacian, dating from about 35,000 to 21,000 years ago, and it shows a higher proportion of prismatic blades than the Aurignacian. Its type fossil is the Gravette stone point, which functioned as a knife, a projectile weapon, or possibly both. The Gravettian is also marked by "laurel leaf" shaped points (chapter 10).

In Western Europe the **Solutrean** follows the Gravettian. Dating from about 24,000 to 18,000 years ago, this industry appears during the coldest time in the last 120,000 years (the Last Glacial Maximum [LGM]; chapter 12). Its type fossil is the shouldered point; the shoulder is used to embed the point in a haft. These points are almost certainly projectile weapons. In Italy and Central/Eastern Europe, there is no Solutrean; instead, an **Epigravettian** industry follows the Gravettian.

The last Upper Paleolithic industry in Western Europe is the **Magdalenian**, dating from about 18,000 to 12,000 years ago, a climatically variable period colder than today but warmer than the LGM (chapter 12). This industry is famous for its cave art and complex tools made from antler and bone (chapter 11) and is contemporary with late Epigravettian industries in Italy and Central and Eastern Europe. These final Paleolithic industries are sometimes followed by a brief "Epipaleolithic" phase, beginning about 11,500 years ago (chapter 12) in which most lithics are smaller versions of Upper Paleolithic tools. The Epipaleolithic grades into the Mesolithic about 10,000 years ago. The Mesolithic has a simplified stone tool kit characterized primarily by small sharp implements known as microliths (chapter 10).

TABLE 2.2 Chronology of archaeological industries and their associated hominins from regions of Europe, about 56 ka cal. BP to about 6 ka cal. BP (dates approximate)

Years cal BP	Iberia/France	Italy	Central Europe	Balkan Peninsula
6 ka	Neolithic (to ca. 4 ka) Modern Humans			
8 ka		Neolithic (to ca. 4 ka) Modern Humans		
10 ka	Mesolithic (to ca. 7 ka) Modern Humans		Neolithic (to ca. 5 ka) Modern Humans	Neolithic (to ca. 5 ka) Modern Humans
12 ka	Azilian / Asturian Modern Humans	Sauveterrian/Mesolithic (to ca. 7 ka) Modern Humans	Mesolithic (to ca. 7 ka) Modern Humans	Epipaleolithic (to ca. 9 ka) Modern Humans
14 ka				
16 ka				
18 ka	Magdalenian (to ca. 12 ka) Modern Humans		Magdalenian (to ca. 12 ka) Modern Humans	
20 ka		Epigravettian (to ca. 11 ka) Modern Humans		
22 ka			Epigravettian (to ca. 12 ka) Modern Humans	
24 ka	Solutrean (to ca. 18 ka) Modern Humans			
26 ka	Proto-Solutrean (to ca. 23 ka) Modern Humans			Epigravettian (to ca. 12 ka) Modern Humans
28 ka				
30 ka				
32 ka	Gravettian (to ca. 26 ka) Modern Humans	Gravettian (to ca. 25 ka) Modern Humans		

34 ka					
36 ka			Gravettian (to ca. 21 ka) Modern Humans	Gravettian (to ca. 25 ka) Modern Humans	
38 ka					
40 ka	Aurignacian (to ca. 29 ka) Modern Humans		IUP: Bohunician/Szeletian Neandertals? Modern Humans?	Aurignacian (to ca. 34 ka) Modern Humans	
42 ka		Aurignacian (to ca. 29 ka) Modern Humans	Aurignacian (to ca. 29 ka) Modern Humans / Protoaurignacian		
44 ka	IUP: Châtelperronian (to ca. 40 ka) Neandertals? / Protoaurignacian (to ca. 40 ka) Modern Humans	Protoaurignacian Modern Humans? / IUP: Uluzzian (to ca. 40 ka)	Modern Humans?	IUP: Uluzzian (to ca. 40 ka)	
46 ka		Neandertals? Modern Humans?		Neandertals? Modern Humans?	IUP: Bachokirian Modern Humans
48 ka					
50 ka					
52 ka	Middle Paleolithic (to ca. 40 ka in parts of Iberia) Neandertals				
54 ka	IUP: Neronian (Rhône Valley) Modern Humans				
56 ka	Middle Paleolithic Neandertals	Middle Paleolithic (to ca. 45 ka) Neandertals	Middle Paleolithic (to ca. 45 ka) Neandertals	Middle Paleolithic (to ca. 45 ka) Neandertals	Middle Paleolithic (to ca. 45 ka) Neandertals

Note: IUP = Initial Upper Paleolithic; ka cal BP = thousands of years before present, calibrated.

appealing to the Law of Superposition: the fact that in layers of sediment the deeper sediments are almost always older than the ones closer to the present-day surface. How much older? It is often hard to tell without having a firm grasp of sedimentation rates at the site in question. Remember Laganne estimating Vézère Valley erosion rates? This is more or less the converse principle.

Knowing that one layer is older than another is useful information, but what many of us want to know is a skeleton's, artifact's, or layer's exact age, and this is the realm of absolute dates. In Classical archaeology, researchers use historical records to determine exact dates. We know, for example, from Pliny the Younger that Mount Vesuvius erupted for two days beginning on August 24, 79 CE, burying towns such as Pompeii and Herculaneum under a deep blanket of ash.[7] Unfortunately, we have no such historical records for the Paleolithic, so we must rely on radiometric dates. With some exceptions, radiometric dates are based on radioactive isotopes and their known rates of radioactive decay. In addition to a known rate of decay, all radiometric "clocks" need a zeroing mechanism—a time at which the clock starts running.

For the Upper Paleolithic, the most widely used radiometric dating method is the first such method developed: radiocarbon, or carbon-14 (^{14}C) dating. All living things are largely composed of organic (carbon) compounds. Plants make these compounds from the carbon dioxide (CO_2) they take in from the air or molecules they absorb through their roots. Animals get their carbon from the plants and animals they eat. When plants and animals die, their remains are broken down, and ultimately much of the carbon in their bodies is again released into the atmosphere as CO_2.

Approximately 99 percent of the carbon on the planet is carbon-12, its most common stable form. However, there are other, heavier, isotopes of carbon, one of which is radiocarbon, or carbon-14. Carbon-14 is made in the atmosphere when the capture of a thermal neutron created by cosmic rays causes the ejection of a proton, altering a nitrogen-14 atom into a carbon-14 isotope. These carbon-14 atoms decay back into nitrogen asymptotically with a half-life of 5,730 years. If you could somehow put a gram of carbon-14 into a lock box (pretty much an impossible task because only about 1 of every 10^{12} carbon atoms is ^{14}C) and come back in 5,730 years, you would only have half a gram left. In another 5,730 years, there would be only a quarter of a gram left, and so on. By opening your box, you would be releasing into the air nitrogen gas that had once been carbon-14.

During life, all living things take in carbon, and the carbon in their tissues has an amount of radiocarbon proportional to the amount of ^{14}C in the atmosphere. The death of the organism is the zeroing event: it is no longer incorporating radiocarbon into its tissues and now begins to lose radiocarbon due to the isotope's radioactive decay. Therefore, like the lock box example, half its original radiocarbon is gone in 5,730 years, half of that is gone 11,460 years after its death, and so on. Remember that only a tiny proportion of the organism's carbon is ^{14}C. Thus, after many millennia, there will not be enough radiocarbon left to detect. As a result of this and the near certainty of modern carbon contamination, there are limits as to how far back in time an organic object can be radiocarbon dated. Regarding contamination, there are pretreatment protocols, such as ultrafiltration for bone, that purify the sample so one is dating only its original collagen. In the absence of pretreatment, most researchers do not trust a radiocarbon age of 40,000 years BP (Before Present [1950]) or older. They assume this to be a minimum age, and that the actual age of the organic material they are dating (e.g., charcoal, bone, antler) could be much older still. As Wood et al. note, 1 percent modern carbon contamination will cause a 50,000-year-old object to appear to be 37,000 years old![8] Following pretreatment protocols can extend the reliable radiocarbon dating range to about 55,000 years BP.

Conventional radiocarbon dating requires the destruction of a fair amount (perhaps 100 grams or 3.5 ounces) of organic material in order to get detectable levels of ^{14}C. Although it is more expensive, most radiocarbon dates today use accelerator mass spectrometry (AMS) to measure the number of ^{14}C atoms in a much smaller sample. I've successfully gotten dates on bone samples weighing 5 grams (0.2 ounce). The dating ceiling remains around 55,000 years BP, however.

There are other caveats I should mention. First, there is error involved in estimating time from the zeroing event; for this reason, all radiocarbon ages are reported with error. For example, Jacobi and Higham generated an AMS radiocarbon date on the "Red Lady" of Paviland (chapter 1).[9] They reported an uncalibrated radiocarbon date of 28,400 ± 320 years BP. This means that there is a 68 percent probability that the actual age of the specimen lies between 28,080 and 28,720 BP. There is also about a 2 percent chance that it is younger than 27,760 BP and a 2 percent chance it is older than 29,040 BP. For me as a paleoanthropologist, this is a nice, precise, date, but let's have a bit of fun with it. Let's say that Columbus landed in the Americas in 1492 ± 320 years. That would mean there was a 68 percent chance that he landed sometime between 1172 and 1812. It also means that there is a 2 percent chance he landed prior to the year 852, and a 2 percent chance that he will land sometime after the year 2132. Precision is in the eyes of the beholder!

Another caveat is that the amount of radiocarbon in the atmosphere is decidedly not stable. Solar activity affects it, and nuclear weapons detonated above ground in the 1940s and 1950s also generated large amounts of ^{14}C. To correct for these differences, scientists created calibration curves that can be used to alter a radiocarbon age to an age in calendar years, indicated by the abbreviation (cal BP).[10] The ways in which radiocarbon ages are calibrated are ingenious. First, in some regions of the world, ancient wood is preserved due to special environmental conditions. In the desert southwest of the United States, wood survives due to the aridity of the environment. In the bogs of northwestern Europe, wood is preserved by being immersed in a cold, acidic, anaerobic environment.

Within a local area, one can seriate tree rings from the youngest (recently dead trees) to the oldest (archaeological) by matching their annual rings. In drought years, their rings are narrower; in wet years, they are wider; and one can line them up like wooden UPC codes. Using dendrochronology, one can pinpoint the exact year a particular tree ring grew more than 11,000 years ago. Acquire an AMS date of that particular ring, compare it to the exact age you know the ring to be, and one starts to build a calibration curve. In Germany and Ireland, dendrochronology now extends back about 13,900 years.

Other annual phenomena have also been used to calibrate radiocarbon dates. In glacial lakes, there is a huge algal bloom during the brief summer. At the end of summer, the dead algae sink to the bottom of the lake, forming discernible annual layers called varves. These layers can be measured from last year's varve down tens of thousands of years, and because they are composed of organic material, they can be radiocarbon dated. As with the tree rings, one dates an annual layer for which one knows the exact year it was laid down, and the curve is created by the comparison between the two.

The calibration curve remains imperfect, however. Throughout prehistory there have been radiocarbon plateaus, in which more radiocarbon was being made in the atmosphere than was being lost through decay. This tends to make objects from those time periods appear younger than they really are, and it greatly increases the error range of their radiocarbon dates. Unfortunately, one such plateau occurs right around the time modern humans appeared to be replacing Neandertals in Europe. From my perspective, this is truly maddening!

Aside from radiocarbon, other absolute dating methods are frequently used in paleoanthropology, especially when a site or layer yields what appears to be a minimum AMS radiocarbon date. Of these methods, Argon-Argon dating, a more accurate form of Potassium-Argon (K-Ar) dating, is the most reliable. Here's how it works. The most common stable form of potassium is ^{39}K. There is a small proportion of potassium that is a radioactive isotope, ^{40}K. It decays to ^{40}Ar with a half-life of 1.248 billion years, meaning it can be used to date very ancient rocks indeed. The substrates that are dated using this method are crystals within volcanic layers called tuffs (in some places these layers are called "members"). Here, the eruption of the volcano is the zeroing mechanism, as any ^{40}Ar

gas present in the rock will escape during the high heat and energy of an eruption. Therefore, any ^{40}Ar found in a volcanic tuff thousands, millions, or even billions of years later must be the product of the natural decay of ^{40}K. In the lab, a single crystal from a volcanic layer is irradiated with neutrons. This converts its ^{40}K to ^{39}Ar, an isotope not found in nature, and unlike potassium, a gas that can serve as potassium's proxy. The proportion of ^{40}Ar to ^{39}Ar is then measured to determine the tuff's age.

When I was in graduate school, it was thought that Argon-Argon dating could only be used on rocks hundreds of thousands of years old or older, but in 1997, Paul Renne and colleagues used the method on ash from Pompeii and generated a date of 72 CE ± 94 years. Recall that the actual date of the eruption was 79 CE, so the method was off by a mere 7 years. Thus Ar-Ar dating can be used on volcanic materials that are just a few thousand years old. The method's only drawback is it can only be employed in areas that are, or were, volcanically active.

Another dating method frequently employed in karstic (limestone cave) environments is Uranium-Series or Uranium-Thorium dating. It is used to date cave formations such as flowstones, stalagmites, or stalactites, known as speleothems. This method is based on the fact that ^{234}U decays to ^{230}Th with a half-life of 245,500 years. Its zeroing mechanism is that uranium is water-soluble and thorium isn't; therefore, in water-lain deposits, or deposits that precipitate out of water such as calcium carbonate flowstones, any thorium present is the result of the radioactive decay of the previously dissolved uranium. A related method, Uranium-Lead (U-Pb) dating, is based on a similar principle involving the more common ^{238}U isotope as it decays through a long and complicated series to ^{206}Pb.

Finally, there is a group of related dating methods that take advantage of the fact that crystals buried in sediments containing radioactive elements will over time trap electrons in their crystalline matrix due to background radiation. The number of trapped electrons is measured, and using a number of assumptions about the underlying level of background radiation, the rise and fall of water levels, and the timing of the uptake of the electrons, an estimated age is generated. The first of these methods is thermoluminescence (TL). TL is often used to date flints that had been burned in a prehistoric hearth, or ancient fired clay. In this case, the heating of the flint or the firing of the clay is the zeroing event, as any earlier trapped electrons in the crystalline structure of these artifacts were released by the heat. What the researcher is then doing is measuring the electrons that have been trapped since it was used in that particular prehistoric context. The sample is heated in the laboratory, releasing the trapped electrons, which are measured.

Recently, optically stimulated luminescence (OSL) methods have become popular. Here one is measuring the time since a crystal was exposed to light: light, like heat, causes trapped electrons to be released. OSL samples must therefore be protected from the light. This is done by driving a metal tube into the profile, or wall, of one's site, and quickly covering the open end of the tube. The tube will contain crystals from the sediment. Crystals within the sample are then tested in the laboratory by exposing them to specific wavelengths of light, releasing and measuring the trapped electrons.

The last of these related techniques is electron spin resonance, or ESR. Relative to the magnetic field, trapped electrons within a crystal have a different spin than the original electrons. The number of trapped electrons can be measured by exposing the sample to microwaves and measuring the sample's resonance signal. In paleoanthropology, this method is most frequently used to date tooth enamel, usually of bovids, cervids, or equids because their enamel is thicker, but it has been successfully employed on hominin teeth as well. One advantage of this technique over TL or OSL is that the trapped electrons are not expelled; therefore, the same tooth can be dated and redated multiple times.

Ages generated by these methods (TL, OSL, ESR) tend to have wide error ranges, but they are useful for sites that are either too old for radiocarbon, lack organic material for radiocarbon, or have no flowstones or volcanic tuffs. These methods can also be used as a check on each other. After all, an archaeologist has much more confidence in the age of a site if multiple methods yield similar results!

3

The Abel to Our Cain?

Homo neanderthalensis

The first time I met the "Old Man" of La Chapelle-aux-Saints was February 28, 1994. It was like meeting a celebrity, someone you've read about for years, only now you get to see this person in the flesh. Well, except for the flesh part. You see, the Old Man is one of the most famous Neandertals of all time (La Chapelle-aux Saints 1, or LC1). He's famous because the person charged with describing the skeleton, Marcellin Boule (whom we met when he was studying the Balzi Rossi fauna), was intent on showing that Neandertals were an evolutionary dead end. Thus he emphasized the Old Man's features as being apelike—comparing his facial prognathism (projection) to that of a chimpanzee, for example, or interpreting the backward tilt at the top of his tibias as indicative of a stooping, bent-kneed gait.[1] Boule's landmark (albeit flawed) work on this skeleton was published from 1911–1913, and its broad acceptance in the early twentieth century is why, if someone calls you a Neandertal, it's usually meant as an insult.

Here is what I wrote in my notes the day I met LC1: "I did the vertebrae first, & the first thing I noticed was that T1 (according to Boule) is not T1 at all, but C7." In other words, Boule misidentified the seventh cervical (or lowest neck) vertebra as T1, the first (or highest) thoracic vertebra. Note that T1 articulates directly beneath C7. I noticed later that same day that Boule's second lumbar vertebra (L2) was really a third lumbar vertebra (L3).

I must have thought this "profound" discovery of mine was important enough to leave a note in the box with the remains for future researchers, because in 2018 the bioanthropologist Chuck Hilton saw it in Paris and sent me a photo of it (figure 3.1)!

The Old Man and I met at the Musée de l'Homme (Museum of Man), located in the Place du Trocadéro, just across the Seine from the Eiffel Tower. Place du

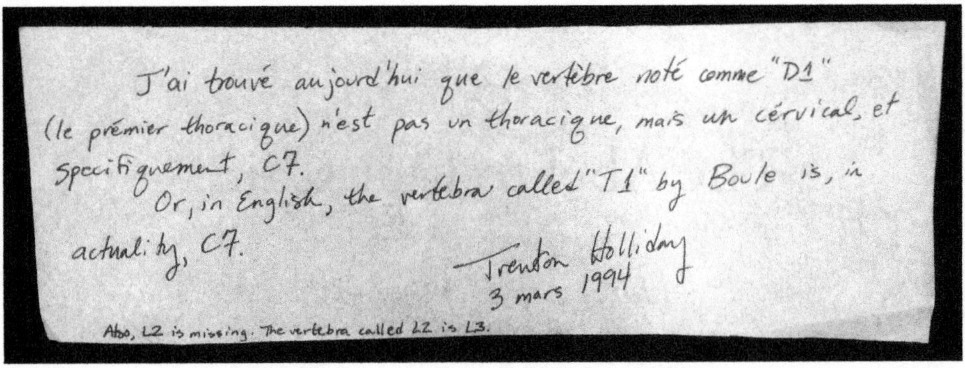

J'ai trouvé aujourd'hui que le vertebre noté comme "D1" (le prémier thoracique) n'est pas un thoracique, mais un cérvical, et specifiquement, C7.

Or, in English, the vertebra called "T1" by Boule is, in actuality, C7.

— Trenton Holliday
3 mars 1994

Also, L2 is missing. The vertebra called L2 is L3.

FIGURE 3.1 The note I left with the skeleton of La Chapelle-aux-Saints 1 in 1994.

Source: Photographed at the Musée de l'Homme in 2018 by, and courtesy of, Charles Hilton.

Trocadéro is upscale and is a great place to run into French celebrities, if you're into that sort of thing. I worked in the museum for several weeks and have fond memories of my time there. My first day I introduced myself to the security guard at the employee entrance—"*Bonjour, mon nom c'est Trenton Holliday; je suis étudiant en doctorat à l'Université du Nouveau Mexique*" ("Hello, my name is Trenton Holliday; I am a doctoral student at the University of New Mexico"). From that moment on he only spoke Spanish to me, which I thought was a kind (albeit somewhat misguided) gesture.

I also have fond memories of Jean-Louis Heim (1937–2018), a paleoanthropologist at the Musée de l'Homme known for his work on the La Ferrassie Neandertals. He was also one of the consultants for the Préhisto Parc mentioned in chapter 1. Jean-Louis was a gregarious bon vivant whose ability to take people away from work for a coffee or wine break was legendary. When he heard that I had grown up in Baton Rouge, he told me "I've been to Baton Rouge! Nothing happens there!" He also told me about being in a restaurant somewhere in south Louisiana and realizing after half an hour that the people at the table next to him were speaking Louisiana French. He said he tried to speak French with them, but he couldn't understand them, nor they him, so they switched to English. I took that as a hint that we should stick to English too.

The last time I saw Jean-Louis was in 2007, when I served on the doctoral thesis committee of one of his PhD students, Mélanie Frelat (now a highly regarded paleoanthropologist in her own right). He was the same as ever—a huge personality. He is sorely missed.

WHO WERE THE NEANDERTALS?

The Neandertals almost certainly evolved in Europe sometime in the Middle Pleistocene, which began 774,000 years ago. There are hints of their morphology in the hominins from the site of Sima de los Huesos, in the Atapuerca Hills region of north-central Spain (see figure 1.1). These remains include more than 6,700 fossil pieces, representing a minimum of twenty-eight individuals dated to 430,000 years ago. Some (e.g., Stringer) argue that these fossils should be referred to *Homo neanderthalensis*, which would make them the oldest Neandertals.[2] Ancient DNA from a femur and an incisor from the site shows that the Sima de los Huesos hominins share many derived alleles with later Neandertals and with Denisovans, another prehistoric hominin population (chapter 6).[3] Thus not only do we have morphological evidence for a link between these hominins and the Neandertals but genetic evidence as well!

The Sima de los Huesos hominins are almost certainly ancestral to the Neandertals, but I, like Juan Luis Arsuaga (the leader of the Atapuerca team), hesitate to put them in *H. neanderthalensis* because they do not show the full suite of derived characters we see in later Neandertals. For me, the earliest Neandertal fossils are the Ehringsdorf Neandertals from central Germany of about 230,000 years ago.

Neandertals expanded beyond Europe about 75,000 years ago. We find their skeletons in the Levant, the Zagros Mountains of Iraq and Iran, eastward to Uzbekistan, and farther east still into southern Siberia near the Altai Mountains. As of this writing, no Neandertals have been found in Africa, China, South or Southeast Asia—nor in Australia or the Americas, continents only colonized by *H. sapiens*. The last of the Neandertals may have been relict populations in southern Iberia who disappeared 30,000 years ago, but this date is controversial (chapter 8).

THERE CAN ONLY BE ONE: AN ABBREVIATED HISTORY OF EUROPEAN NEANDERTAL DISCOVERIES

No book on Cro-Magnons can be written without referencing Neandertals. The two hominins' histories are inextricably linked, having both been discovered in Western Europe in the mid-nineteenth century. Here we have two populations that are clearly anatomically different from each other. Although most

Neandertals antedate the Cro-Magnons by millennia, the two groups coexisted in Europe around 40,000 years ago, and as such, just like *Highlander*, there can only be one (primary ancestor of us, that is). Cro-Magnons have long been represented in the literature as the more successful, smarter, and ultimately triumphant species (us), and Neandertals became the brutish "other"—the evolutionary dead end that couldn't compete with us.[4] I have long argued that this is an overly simplistic view of the relationship between these two hominins.

In the Introduction I mentioned that the earliest Neandertal discoveries were not immediately recognized as being either prehistoric or different from *Homo sapiens*. In Engis, Belgium, around 1829 or 1830, the first Neandertal fossil, a child's skull, was uncovered by the physician Phillipe-Charles Schmerling (1790–1836), who failed to see its taxonomic significance. Unfortunately for Schmerling, Charles Lyell visited the site in 1833 and remained unconvinced of the antiquity of the Engis remains. Then, in 1848, a Neandertal skull enigmatically appeared out of Forbes' Quarry in Gibraltar. It made its way to London, where it was only belatedly recognized as a Neandertal. What if either of these finds had been immediately recognized as prehistoric nonmodern humans? Imagine a world in which people insult each other by throwing out the epithet "Engisian" or "Gibraltarian" instead of "Neandertal"—it could have happened!

But alas, that was not to be because the first recognition that there was a prehistoric human different from people today rests on a different cave, in the Neander Valley of western Germany near Düsseldorf. The cave was the *Kleine Feldhofer Grotte* (Little Feldhofer Cave). It was named for a nearby farm, and it *was* small—about the size of my daughter's attic bedroom. According to Schmitz et al., it was but 3 meters (10 feet) in width and height, and only 5 meters (16 feet) deep. Its entrance was smaller still—less than 1 meter (3 feet) wide. The cave was also about 20 meters (65 feet) above the nineteenth-century valley floor.[5] You'll notice that I've used the past tense with this cave because it was completely destroyed, as were the limestone cliffs in the valley in which it was found. Today the Neanderthal Museum,[6] which opened in 1996, is located on a flat spot of ground a short distance from where the little cave once stood.

As Schmitz et al. relate, by the mid-1850s, the demand for limestone for the Prussian construction industry had reached this tiny valley.[7] In 1854 Wilhelm Beckerschoff and Friederich Wilhelm Pieper founded a quarrying company there to help supply limestone to new construction projects. Before quarrying in earnest could begin, however, workers had to remove sediments from the caves to reduce impurities in the quarried limestone. In so doing, they often encountered bones, particularly those of cave bears.[8] A schoolteacher and avid natural historian from the nearby town of Elberfeld, Johann Fuhlrott (1804–1877), had arranged with quarry workers to set aside any fossils they might happen to find. When the foreman informed him in August 1856 that they had uncovered and set aside the skeleton of a cave bear, Fuhlrott wanted to retrieve it, but he was unable to do so until a couple of weeks later. Upon arrival, he immediately recognized that the skeleton (a calotte, or skullcap, and most of the major limb bones) was not that of a cave bear but rather of a human, albeit one unlike anyone alive today. He asked where the skeleton had been found, and they told him in the Little Feldhofer Cave. He asked if there had been artifacts associated with it (no one seemed to think there were)—he even went to the site to look for more remains, but it was all in vain.[9]

Fuhlrott needed an expert to verify that the bones were those of a prehistoric human, so he brought them to the attention of an anatomy professor at the University of Bonn, Hermann Schaafhausen (1816–1893). This was fortuitous, and three years before publication of Darwin's *On the Origin of Species* Schaafhausen agreed that the remains were of a "Pre-Celtic" race of people and that the specimen's strange anatomy was *not* due to any known pathological or disease process.

In 1857 they jointly presented their findings at a scientific conference in Bonn, but once again I must remind you that scientists are a skeptical lot—few, if any, German scholars agreed that the Feldhofer individual was a prehistoric human. The most creative critique of the specimen's antiquity came from one of Schaafhausen's own colleagues, August Mayer (1787–1865). Mayer interpreted the bowing of the specimen's femora (thighbones) as due to a life in the saddle coupled with a childhood case of rickets. He opined that the remains were likely those of a Cossack soldier from Russian forces commanded by Alexander Tchernyshov (1786–1857)—an army that fought Napoleon in Germany and France from 1813 to 1815. Mayer suggested that, upon being mortally wounded, the soldier had crawled into the Feldhofer Cave and died.

This argument was later ridiculed by the man known as "Darwin's Bulldog," Thomas Henry Huxley (1825–1895), who wanted to know how the mortally wounded man had managed to get covered in deep sediments before (or soon after) dying. But the rickets portion of Mayer's explanation was accepted by the most vocal opponent of evolutionary theory in Germany—the Prussian physician Rudolf Virchow (1821–1902), an expert on the impact of disease on human tissues and considered by many to be the "Father of Pathology." Although I doubt it would have made much of a difference, given Virchow's prominence and his vehement opposition to evolution, Schmitz et al. cannot help but wonder how different the reception by the German scientific community of Feldhofer might have been had the specimen been discovered in a controlled excavation alongside the Mousterian tools that had also been there.[10]

More Neandertal remains would soon come to light. In 1864, the British surgeon and paleontologist George Busk (1807–1886) and the Scottish geologist Hugh Falconer (1808–1865) presented a paper on the Forbes' Quarry (Gibraltar) cranium, recognizing its similarities to Feldhofer. In 1866, the Belgian geologist Édouard Dupont excavated a large cave called Le Trou de la Naulette not far from the French border in Belgium's Namur province, recovering a human mandible, ulna, and metacarpal. These bones were in a level that lay beneath four stalagmitic layers and came from the same level as those of woolly rhinoceroses, mammoths, and reindeer, suggesting an ancient date. Furthermore, the mandible lacked a chin, prompting Broca to call it the most primitive human mandible known. The specimen was not at first recognized as a Neandertal, primarily because neither Feldhofer nor Forbes' Quarry had mandibles. Virchow examined la Naulette and claimed the lack of a chin was due to a pathological lesion at the front of the mandible; there is, however, no evidence of any type of bony lesion on the specimen.

In 1880 at Šipka Cave in northern Moravia (then part of the Austro-Hungarian Empire), the Czech archaeologist Karel Maška (1851–1916) uncovered the mandible of a child in a layer yielding both Mousterian tools and extinct animals. Only the very front of the mandible was preserved. One adult central incisor was missing postmortem, but the other had erupted, as had both adult lateral incisors. Based on the mix of adult and "baby" teeth present, the specimen was likely eight to ten years old when he or she died. Virchow remained unswayed, saying that the mandible was far too thick to be that of a child.[11] Instead, it was that of a pathological adult, most of whose adult teeth had failed to erupt due to an unknown disease process.

The find that would cause most scientists to abandon Virchow's pronouncements like these came in 1886 at the site of Spy d'Orneau, a cave in Belgium's Namur province. A lawyer and an amateur archaeologist, Marcel de Puydt, and a fish paleontologist from the University of Liège, Marie Joseph Maximin ("Max") Lohest, began working in the cave in July of that year. Spy d'Orneau had been known for many years to yield fossil bones, and as a result, many areas of the cave had been disturbed by amateur excavators. However, de Puydt and Lohest found a spot near the front of the cave that appeared undisturbed. There, underneath a stalagmitic layer, they found not one but two adult Neandertal skeletons associated with Mousterian artifacts and extinct animals.

Lohest felt a bit out of his league with human fossils, so he brought in the anatomist Julien Fraipont to help him and de Puydt study the Spy specimens. Together they worked feverishly and were able to produce a monograph on the skeletons in 1887.[12] The Spy crania were long and low, with large brow ridges reminiscent of those of Feldhofer and Forbes' Quarry. In fact, the cranial shape of both Spy crania was so similar to Feldhofer that denying their morphological affinity was untenable. The mandible of Spy 1 was thick and lacked a chin, like that of La Naulette. Here we have not one but two individuals, recovered with Mousterian tools, found beneath an undisturbed stalagmitic layer and associated with extinct animals. They simply cannot be pathological modern humans. But if Neandertals were prehistoric humans who made and used Mousterian tools, were they contemporary with Cro-Magnons? The archaeological record suggested that Neandertals antedated Cro-Magnons because Mousterian layers were always found below those of the Upper Paleolithic. Were Neandertals then ancestral to Cro-Magnons? In the nineteenth and early twentieth centuries, many scholars asked this question would give a resounding "No!" in response. We'll talk about this in just a bit; I first want to discuss two final Neandertal sites from southwestern France.

In 1908, two brothers (and priests), Amédée and Jean Bouyssonie, were excavating a small cave near the village of La Chapelle-aux-Saints, in the Corrèze region of France, about 56 kilometers (35 miles) east-northeast of Les Eyzies, when they uncovered an intentionally buried male Neandertal skeleton[13]—the one from the beginning of this chapter. He had been lain on his back in a shallow pit in the fetal position, his knees up near his abdomen. His long, low, skull had big brows, and he was missing most of his teeth (figure 3.2). For this reason, like Cro-Magnon 1, he would become known as the *vieillard*, or "Old Man." However, unlike Cro-Magnon 1, the teeth of LC1 had been lost years before his death, based on the large amount of alveolar bone resorption in both his mandible and maxillae.[14]

For advice on the find, the Bouyssonie brothers contacted the prehistorian Abbé Henri Breuil (1877–1961). Breuil was a Jesuit priest, an expert on Paleolithic art, and a talented artist in his own right who by 1908 was well on his way to becoming world famous. He had dug at Cro-Magnon while he was a seminary student in 1897 and 1898, but more important, in 1901, while working with the physician and prehistorian Louis Capitan (1854–1929) in Les Eyzies, he had discovered hundreds of examples of Paleolithic cave art at the sites of Les Combarelles and Font-de-Gaume. Breuil and Capitan grouped this art with stylistically similar paintings from the Spanish Altamira Cave, discovered in 1879, but which had been dismissed by prehistorians as beyond the artistic capabilities of Pleistocene humans. In fact, two years prior (1906), Breuil and his former professor Émile Cartailhac had published a book on the art from Altamira that included beautiful color reproductions rendered by Breuil himself.

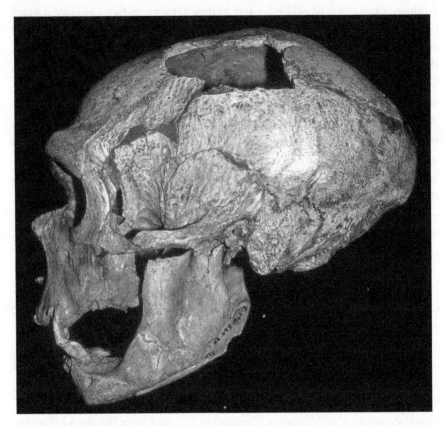

FIGURE 3.2   Left lateral view of La Chapelle-aux-Saints 1 Neandertal skull.

Source: Courtesy of Erik Trinkaus.

Jean (the younger of the Bouyssonie brothers) and Breuil had been in seminary together and remained lifelong friends. When the brothers asked their friend to whom they should send the fossil for anatomical analysis, Breuil recommended Marcellin Boule at the Muséum National d'Histoire Naturelle in Paris. Boule and Breuil knew each other from their student days at the University of Toulouse, as both had studied under Cartailhac. (To me, this story is evidence that it's not just *what* you know but *whom* you know that helps determine the course of your life.)

Recall from chapter 1 that Boule was trained in paleontology, which was why he had been brought in to study the fauna from the Grimaldi Caves. He wanted to make human paleontology compatible with the rest of the discipline, and as

a paleontologist he held fast to the (true) idea that most fossil species are evolutionary dead-ends that leave no descendants. In 1905 he had argued that the
deepest skeletons in the Grotte des Enfants were associated with warm climate
fauna such as *Rhinoceros mercki*, *Elephas antiquus*, and hippopotamuses—fauna
he considered indicative of the early Pleistocene—so he was convinced that
Neandertals and early *Homo sapiens* were penecontemporary (living at about the
same time) and, therefore, only the latter could be ancestral to us. The current
cultural milieu was also one in which notions of inevitable progress held great
sway. Viewed through this lens, Boule's overemphasis on the so-called primitive
and backward features of La Chapelle-aux-Saints makes sense, and it is this view
of the Neandertal as the primitive "other" that has pervaded paleoanthropology
ever since.

More Neandertal finds were soon on the way. In terms of associated skeletons,
no Neandertal site in Europe beats La Ferrassie, with Denis Peyrony (1869–1954)
and Louis Capitan ultimately recovering the remains of at least seven Neandertals (men, women, and children) there. The rock shelter at La Ferrassie is in the
valley of a small unnamed tributary to the Vézère some 6.25 kilometers (3.9 miles)
west-northwest from Les Eyzies and 3.7 kilometers (2.3 miles) north-northeast
from the Vézère River town of Le Bugue. It has a wonderfully deep archaeological sequence, ranging from the Gravettian at its top through the Aurignacian,
Châtelperronian, and Mousterian.[15] Peyrony and Capitan began exploration of
the site in 1899 in the Upper Paleolithic levels and worked their way down over
several seasons. Ten years later, in 1909, near the base of the Mousterian levels
and close to the back wall of the shelter, they discovered the nearly complete
skeleton of an adult male Neandertal (La Ferrassie 1, or LF1). His body had been
laid on the surface in a supine position, with hips and knees flexed, his left arm
at his side, and his right arm flexed. In 1910, they found a second Neandertal,
a presumed female (La Ferrassie 2, LF2). She had been positioned on her right
side with her hips and knees flexed, and her arms resting on her knees. The
long axis of her body was aligned with that of LF1; the crowns of their heads
were facing each other, separated by only half a meter, yet another example of
intentional burial.

The remaining partial skeletons, recovered later, were children buried in pits.
LF3, represented by head, upper limb, and foot bones, is now thought to be
the same individual as the LF7 talus (ankle bone) and considered to have been
around ten years old at the time of death. LF4b, a neonatal skeleton, is represented by preserved parts of the cranium, thorax, and limbs. LF5, a fetal skeleton,
included part of the cranium, as well as both humeri, femora, and tibiae. LF6
is a largely complete skeleton of a three- to five-year-old child, which is said to

have been buried in a flexed position underneath a large slab that functioned as a tombstone. Finally, LF8, estimated to be a child of around two years of age, is represented by a preserved partial cranium, multiple ribs, several vertebrae, and bones of the pelvis and hands.

VOICI LA DIFFÉRENCE: ANATOMICAL FEATURES OF NEANDERTALS VS. CRO-MAGNONS

Many morphological features distinguish us (and the Cro-Magnons) from Neandertals. Most are craniofacial because this is the most informative anatomical region for assigning fossil hominins to species. Many features that distinguish Neandertals from *Homo sapiens* are plesiomorphic characters. In the field of phylogenetics (reconstructing evolutionary relationships among taxa), *plesiomorphic* characters are ancestral or (I hate this word) "primitive" characters. These are contrasted with *apomorphic* (derived) or more recently "evolved" characters. Because evolution is a temporal phenomenon, whether a character is ancestral or derived depends on the scale at which you look. For example, among primates, an opposable thumb is an ancestral character present in the last common ancestor of all primates, and it is retained in almost all primate species today. In contrast, across mammals an opposable thumb would be a derived character that could be used to distinguish primates from other mammals.

The decision as to whether a character is ancestral or derived (its *polarity*) is critically important in phylogenetic (evolutionary) analysis. Why is this the case? In my opinion, the most important contribution to evolutionary biology by the architect of phylogenetic systematics, the German entomologist Willi Hennig (1913–1976), is his assertion that shared plesiomorphic (ancestral) characters are not informative as to whether a group of species have a recent common ancestor. Instead, only derived characters shared among two or more taxa are phylogenetically informative. The Harvard paleoanthropologist Daniel Lieberman used the following intuitive example to illustrate this point. Pentadactyly (having five digits at the end of each limb) is an ancestral character dating back to the first amphibian-like animals that colonized land during the late Devonian period, about 380 million years ago. If we use this feature to determine who is more closely related to whom among a human, a horse, and a frog, it will link the human with the frog to the exclusion of the horse. Based on many other features, we know that the human and the horse are more closely related to each other than either is to a frog. For this reason, ancestral features should not be used as

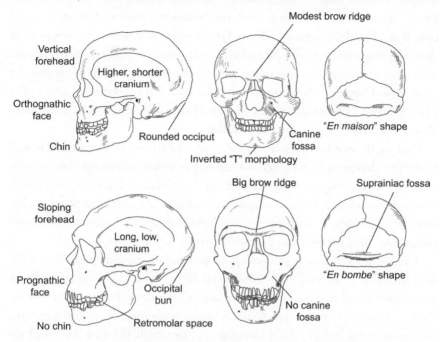

Grotte des Enfants 4 (*Homo sapiens*)

Modest brow ridge

Vertical forehead

Higher, shorter cranium

Orthognathic face

"*En maison*" shape

Chin

Rounded occiput

Canine fossa

Inverted "T" morphology

Big brow ridge

Suprainiac fossa

Sloping forehead

Long, low, cranium

"*En bombe*" shape

Prognathic face

Occipital bun

No chin

Retromolar space

No canine fossa

La Ferrassie 1 (*Homo neanderthalensis*)

FIGURE 3.3 Cranial differences between *Homo sapiens* and *H. neanderthalensis*. Top row, Grotte des Enfants 4 (*H. sapiens*); Bottom row, La Ferrassie 1 (*H. neanderthalensis*). Left to right are left lateral, anterior, and posterior views. Note that both specimens show asymmetries due to postmortem distortion.

Source: Drawings by author.

grouping criteria—they cannot be used to determine evolutionary relationships among taxa because they can lead to false conclusions.

Nonetheless, the presence of an ancestral character state in Neandertals, when contrasted with a derived character state in Cro-Magnons, is of interest. Some of these informative distinctions are depicted in figure 3.3, which compares the skull of a Cro-Magnon (Grotte des Enfants 4) with that of a Neandertal (La Ferrassie 1). First, Neandertals have a long, low neurocranium (braincase) shaped like an American football. This cranial shape is found in most earlier fossil hominins as well, and as such is ancestral for the genus *Homo*. When viewed from the side (lateral view), the modern human cranium is more globular, i.e., shorter from front to back, with a higher, more vertical forehead and a more rounded rear, or occiput. Modern human cranial anatomy is therefore derived. Perhaps counter-intuitively, when viewed from behind, the Neandertal cranium is more rounded than that of modern humans! We use French terms to describe the shape of the skull in posterior view—the Neandertal cranium is *en bombe* (bomb-shaped), whereas the modern human cranium is *en maison* (house-shaped). Note that the back of the Grotte des Enfants 4 cranium looks as if it has walls and a roof (like a house drawn by a child), whereas the rear of the La Ferrassie 1 cranium is more akin to the back of a 1962 Airstream trailer.

Despite these differences in cranial shape, in terms of absolute brain size, both Neandertals and Cro-Magnons had bigger brains on average than do peo-ple today. Correcting for body size by using the encephalization quotient (EQ), however, Neandertals are less encephalized (smaller-brained) than people today, whereas Cro-Magnons are more encephalized than we are. I approach cognitive conclusions drawn from these data with extreme caution (chapter 7).

The Neandertal face is projecting, or prognathic, especially in the middle of the face (i.e., the nasal region and front teeth). This is known as midfacial prognathism, and it is one of the most salient Neandertal features. In contrast, the Cro-Magnon face, like ours today, is largely nonprojecting, or orthognathic, instead showing neuro-orbital convergence, in which the face is tucked up underneath the braincase—not positioned out in front of it. A related feature that distinguishes Neandertals from us is our canine fossa. A canine fossa is a ver-tical furrow of our upper jaw bones close to the roots of our canines. The paleo-anthropologists Scott Maddux and Bob Franciscus have found that this feature tends to form in nonprojecting faces. As such, it is found in all people today but is much less common in nonmodern hominins. It is occasionally seen in archaic hominins with nonprognathic faces, such as the tiny Liang Bua 1 specimen (the "Hobbit"), the holotype (type specimen) of *Homo floresiensis*, an insular species that lived on the Indonesian island of Flores around 60,000 years ago. Addi-tional features distinguish Neandertal from modern human crania; for the inter-ested reader, discussion of these is found in box 3.1.

The postcranial (i.e., below the neck) skeleton of Neandertals also shows key differences from Cro-Magnons. Unlike the Cro-Magnons, who show long dis-tal limbs (forearms and calves), European Neandertals have short forearms and calves (chapter 5). Neandertals also show an ancestrally broad pelvis; in contrast, modern humans have the derived condition of narrower pelves. Neandertals had a larger thorax (ribcage) than Cro-Magnons, which could be linked to oxy-gen demands for powering their larger bodies (chapter 7). There are also subtle anatomical differences between Neandertals and modern humans in terms of muscle size and muscle attachment areas; one example, found along the edge of the shoulder blade, is discussed in chapter 9.

The contrasts between Neandertals and Cro-Magnons are marked and easily quantified. In that sense, having Neandertals serve as "the other" is convenient. However, things become a bit sticky if we try to circumscribe *H. sapiens* anatomically. For the sake of argument, let's just assume for the moment that Neandertals warrant assignation to their own species, in which case *H. neanderthalensis* is the oldest name and therefore has taxonomic priority.[16] If we exclude the Neandertals from our species, how do we define *H. sapiens* anatomically, and can we do it quantitatively? The next chapter explores this thorny question after first introducing the earliest modern humans from Africa and the Levant who are presumed to be the primary direct ancestors of the Cro-Magnons.

BOX 3.1. ADDITIONAL CRANIOFACIAL DIFFERENCES BETWEEN NEANDERTALS AND CRO-MAGNONS

Neandertals have larger teeth than Cro-Magnons. In the anterior teeth, Neandertals show the same large incisors we find in *Homo erectus* 1.5 million years ago, teeth that lie outside the recent human range of variation. Despite these huge incisors, by the time Neandertals reach their thirties, the crowns of their front teeth are worn down to nubs from using them as tools. With regard to posterior teeth, Neandertal premolars and molars are smaller than those of *H. erectus*, but bigger on average than those of *H. sapiens*, albeit within the recent human range of variation. The Neandertal ramus, the part of the mandible that rises to meet the skull's temporomandibular joint and on which the major muscles of mastication (chewing) are attached, is generally smaller than those of their presumed ancestors, but bigger than those of the Cro-Magnons. In terms of these two related features, Neandertals are derived but not to the same degree as *H. sapiens*.

Yet another Neandertal feature that reflects these two shifts (maintenance of a projecting face, but with smaller back teeth and chewing muscles) was demonstrated by Robert Franciscus and Erik Trinkaus: the retromolar space. Like their probable ancestors, Neandertal anterior teeth are in an anterior (or prognathic) position, but within the context of both a shorter tooth row (smaller back teeth) and a narrower ramus (reduced chewing muscle size). Thus, in Neandertals there is a space (the retromolar space) between the back end of the lower third molar, or wisdom tooth, and the front edge of the ramus (see figure 3.3). A retromolar space is present in all Neandertal mandibles, but it is only occasionally seen in other hominins.

Another ancestral feature of Neandertal mandibles is that they, like those of other nonmodern fossil hominins, lack a forward-projecting chin, or mental eminence.

(*continued on next page*)

(*continued from previous page*)

The chin is unique to *Homo sapiens*. It is frequently characterized by an inverted "T" configuration in anterior view, although this specific morphology is not always present in people today.

Neandertals have bigger brow ridges than the Cro-Magnons. Here, too, Neandertals show a mix of ancestral and derived morphology. Many earlier hominins, such as *H. erectus*, are characterized by a single continuous brow ridge that is almost as thick at its lateral margins (the sides of the face) as it is medially (near midline). In Neandertals, the brow ridge remains a continuous ridge of bone (a true "supraorbital torus"), but it is thickest near midline and tapers laterally. In this regard, despite its larger size, the Neandertal brow is derived toward what we see in *Homo sapiens*. Very few people alive today have a true supraorbital torus. Instead, each brow is split by a sulcus (linear depression) into lateral (supraorbital) and medial (superciliary) segments (chapter 4). As in Neandertals, in people who have thick brows today, their brows are thickest near midline and much thinner laterally.

The occipital bone at the back of the cranium also shows some uniquely Neandertal features. First, when viewed from the side, Neandertal crania tend to show an occipital "bun" (*chignon* in French), looking almost as if someone has pinched the back of their heads from above and below. Second, in rear view, just above the attachment point for the posterior muscles of the neck, Neandertals have a short, wide, ovoid depression called the suprainiac fossa (see figure 3.3). Neither of these occipital features appears functional, but both are found in high frequency among Neandertals and are nearly absent in other fossil hominins.

In contrast to Neandertals, the back of Cro-Magnon crania tend to be more rounded in lateral view. In both Cro-Magnons and people today, this smoothly rounded posterior profile may be broken up by the presence of an external occipital protuberance, a small projection of bone at midline just above the attachment of the neck muscles and seen more frequently in males than in females. This protuberance is variably present in fossil hominins, but Neandertals lack it.

4

Fossil and Recent *Homo sapiens*

When I was a graduate student, there was a beautifully rendered polyurethane resin cast of the Qafzeh 9 early modern human skull in Erik Trinkaus's lab at the University of New Mexico. It now resides in my own lab at Tulane (figure 4.1). It was made by the late great casting technician Mario Chech at the Musée de l'Homme in Paris. Chech was truly a master at making achingly realistic research-quality casts of fossil hominins that were also works of art. As was the case with La Chapelle-aux-Saints, I felt like I knew Qafzeh 9 well; she was another old friend.

Likely a female, she was a young adult or late adolescent when she died; the roots of her wisdom teeth had not yet finished their growth, and some of her epiphyseal lines were still visible.[1] Anatomically, she is modern in appearance, with a rounded cranial vault, a small nonprojecting face, a high forehead, and a forward-projecting chin. As a graduate student instructor, I would tell students that if you took the Qafzeh 9 skull and placed it in a Civil War era grave no historical archaeologist would blink an eye. This is amazing because multiple absolute dating techniques suggest that she is somewhere between 94,000 and 115,000 years old!

Qafzeh is in Galilee, close to the city of Nazareth, and the young woman we call Qafzeh 9 was interred in an intentional double burial with the skeleton of a six-year-old child (Qafzeh 10). It is *possible* that these two individuals were mother and child. They are associated with multiple red ochre fragments and were recovered from a Mousterian level in 1967 by a joint Israeli-French team.

I studied the original specimen on March 8, 1994, at the Rockefeller Museum, just outside the walls of Old Jerusalem. After a few minutes with her skeleton, I realized I would have to modify my story about her. Don't get me wrong; her cranium *does* look modern—that's not the problem. The issue is that her bones

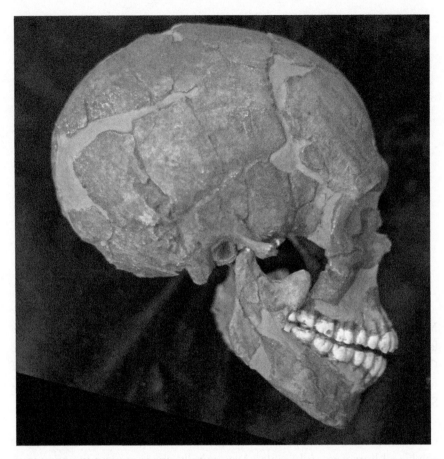

FIGURE 4.1 The cast of the skull of Qafzeh 9 made by Mario Chech, right lateral view.

Source: Photo by author.

are *extremely* fossilized. Picking up her cranium or femur is like picking up a heavy rock. Contrary to what I had been telling my students, a skull as dense as a rock would certainly give pause to any archaeologist digging a Civil War era site. Fossilization is a highly variable process, and in this case, the organic component of bone, its osteoid, most of which is the protein collagen, has been replaced by minerals. The fossilization process occurs at different rates in different burial environments, but I can't think of a single environment in which a skull would become as mineralized as hers in only 150 years!

The French paleoanthropologist Bernard Vandermeersch excavated the Qafzeh 9 and 10 burial and led the bioanthropological studies of all the Qafzeh hominins. In papers in the 1980s and 1990s, he frequently referred to these speci-mens as Proto-Cro-Magnons—reflective of the fact that if one were looking for a population ancestral to the Cro-Magnons these early modern humans, recovered from Mousterian contexts, are among the better candidates. But this all begs the questions: Who are the earliest modern humans, how different are they from us, and where are they found?

AFRICAN ORIGINS OF *HOMO SAPIENS*

The tribe Hominini originated in Africa around 8 million years ago. We have no fossil or archaeological evidence for hominins outside of Africa until about 2.1 million years ago, in the form of stone tools found in China.[2] Recall the Hominin Road Trip we took from the Empire State Building to the Hollywood Walk of Fame (chapter 3). For that trip, we'd reach the 2.1-million-year point in McCartys, New Mexico, a small community of around fifty people in Acoma Pueblo, about an hour's drive west of Albuquerque. For roughly three-quarters of our time as hominins, then, we were an exclusively African taxon. Even after hominins expanded into Eurasia, however, Africa remained home to most hominins throughout the Pleistocene. This is because much of Eurasia was either covered in glaciers or was an inhospitable cold steppic or tundra environment—one in which advanced technology was needed for hominin survival. Perhaps it's no surprise then that Africa is where we encounter the first anatomically modern (or nearly anatomically modern) humans—members of our own species, *Homo sapiens*.

Just as our species varies over distance today, we varied across deep time as well. We cannot expect all the earliest members of our species to look like Qafzeh 9, which in the absence of being fossilized resembles a burial from the nineteenth century. Instead, we might expect the earliest members of our species to be char-acterized by more plesiomorphic (ancestral) characters than are people today. Just as some hesitate to put the Sima de los Huesos fossils in *Homo neandertha-lensis*, not all scholars agree that the specimens I am about to discuss belong in *H. sapiens*. Note, too, that the following discussion is not exhaustive; I cover only the most important fossils.

Currently, the earliest candidates for the title of oldest members of our species come from the site of Jebel Irhoud in Morocco. Jebel Irhoud is a cave located about 48 kilometers (30 miles) inland from the Atlantic, in the foothills of the Atlas Mountains, about 89 kilometers (55 miles) west-northwest of Marrakesh. The first specimen recovered there, Jebel Irhoud 1 (Ir 1), is a nearly complete cranium accidentally discovered by a miner in 1960 or 1961. This brought the site to the attention of archaeologists, who worked there in 1961 and 1962. That work resulted in the recovery of multiple individuals associated with Middle Paleolithic tools. Excavations began anew in 2004, unearthing even more fossil hominins and tools, but perhaps more important, this work provided new radiometric dates (TL and ESR), suggesting that all the hominins from Jebel Irhoud are around 300,000 years old.[3]

The Jebel Irhoud 1 specimen is not terribly modern in its morphology (figure 4.2). Its braincase (neurocranium) is long and low, with an endocranial capacity (1,305 cubic centimeters) just below the recent human mean. The back of the cranium (occiput) is not rounded. It also has a thick, continuous, brow ridge (supraorbital torus). What is modern-looking about it is the size and orientation of its face—it has a much reduced and nonprojecting facial skeleton with a canine fossa. Just how modern is the Jebel Irhoud face? A 3D morphometric

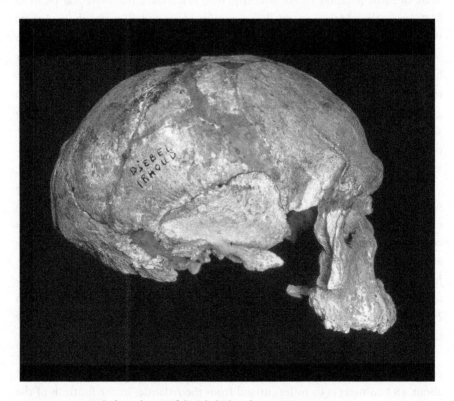

FIGURE 4.2 Right lateral view of the Jebel Irhoud 1 cranium.

Source: Photo courtesy of Erik Trinkaus.

analysis for the face of Irhoud 1 found that it falls among the size and shape variation of recent humans.[4] In terms of its braincase, however, it clusters with archaic members of the genus *Homo*, away from us. This mix of archaic and modern traits led the paleoanthropologist Chris Stringer to consider the specimen *H. sapiens*—but not anatomically modern.[5]

What about the other fossil hominins from Jebel Irhoud? Irhoud 2 (Ir 2) is a neurocranium missing most of its face. Only brow ridges and the top of its nasal bones were preserved. Its cranial capacity of 1,400 cubic centimeters is above the recent human average. Its brow ridges are not continuous like those of Ir 1; rather, they are divided by a sulcus into supraorbital and superciliary segments such as we see among people today (see box 3.1). However, as was the case with Ir 1, its occiput is not rounded, and its braincase shape falls among more archaic members of the genus *Homo* and outside of the range of people today.[6]

Two important finds from the recent excavations at Jebel Irhoud include a partial facial skeleton (Ir 10) and a nearly complete mandible (Ir 11). The Ir 10 fragmentary facial skeleton is more strongly built than that of Ir 1, but nine different 3D reconstructions of its morphology cluster it among recent humans.[7] In contrast, the Ir 11 mandible is not modern in its morphology; it lacks a chin and is considerably larger than the mandibles of people today.

Given all the ancestral characters present in the Irhoud sample, not everyone agrees they should be referred to as *H. sapiens*. However, given its age (300,000 years old), the fact that the face of Ir 1 falls among those of people today is astonishing to me. I am therefore comfortable referring it to our species, recognizing, of course, that people from 300,000 years ago should not be expected to fit completely within the narrow morphological range of people alive today.

Our next potentially early *H. sapiens* hails from Laetoli, Tanzania, a site more famous for the 3.6-million-year-old type specimen of *Australopithecus afarensis* ("Lucy's" species) and for fossilized footprints of roughly the same age than it is for its Middle Pleistocene *Homo* remains. The museum accession number for this Middle Pleistocene specimen is Laetoli Hominid 18 (LH 18). It was discovered by Edward Kandini, a member of Mary Leakey's team in 1976, on the surface of the Ngaloba Beds; for this reason it is often called Ngaloba. In 1993, the Tanzanian geologist Paul Manega generated amino acid racemization dates on ostrich shells from the Ngaloba beds of 205 ± 17 ka and 290 ± 25 ka (ka = thousands of years ago).[8] The older of these dates would make Ngaloba penecontemporary with Irhoud; the younger makes it 100,000 years later than Irhoud.

Ngaloba (LH 18) has a preserved braincase and partial facial skeleton; unfortunately, no bony contact can be made between the two parts (figure 4.3). The braincase is not strikingly modern. Its cranial capacity (1,200 cubic centimeters)

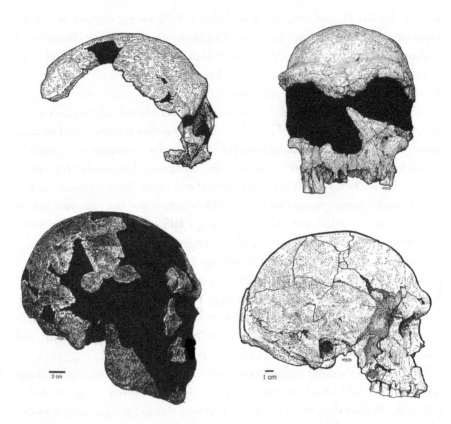

FIGURE 4.3 Four early *H. sapiens* crania from Africa. Clockwise from upper left, right lateral view of the Florisbad partial cranium, anterior view of the Ngaloba (LH 18) cranium, right lateral view of the Herto 1 cranium, and right lateral view of the Omo Kibish 1 skull.

Source: Drawings © Matt Cartmill, used with permission from Matt Cartmill and Fred H. Smith, *The Human Lineage*, 2nd ed. (Hoboken, NJ: Wiley, 2022).

is within the recent human range of variation, but below average. It has a supra-orbital torus, and similar to recent humans, a sulcus separates the brow's medial and lateral components. Its cranial vault is higher than that of either Jebel Irhoud 1 or 2, and its occiput is more rounded than those specimens as well. Hublin et al. found that the neurocranial shape of LH 18 falls just outside the range of recent humans, lying closer to us in morphology than does either Irhoud 1 or 2.[9]

The most modern aspect of LH 18 is its face, which is reduced in size and has a canine fossa. Similar to Jebel Irhoud 1, we have a somewhat archaic-looking cranial vault associated with a more modern-looking face. Unfortunately, the LH 18 face was not complete enough to be included in Hublin and colleagues' analyses.

Florisbad 1 is a partial cranium about the same age as Ngaloba; in 1996, Rainer Grün and colleagues reported an ESR date on the specimen's teeth of 259,000 ± 35,000 BP. Discovered in 1932 in South Africa by T. F. Dreyer, it is the holotype of *Homo helmei*, a species accepted by some paleoanthropologists. Its frontal, parietals, and much of its anterior-most facial skeleton are preserved (figure 4.3). The forehead seems vertical, but it is difficult to know exactly how vertical it is because neither the base of the cranium, the external auditory meatus (ear canal), nor the occlusal (bite) plane of the teeth was preserved. This makes it impossible to orient the cranium in anatomical position. Florisbad 1's cranial vault bones are thick, as is its supraorbital torus, although some argue that the brow hints at division into medial and lateral segments like those seen in modern humans. The most modern aspect of the cranium is its face, which is joined to the neurocranium at midline, is small and nonprognathic, and has a canine fossa. The specimen was not included in the Hublin et al. analyses.[10] All in all, because so little of it is preserved, Florisbad 1 remains enigmatic.

Our next early *H. sapiens* candidate is Omo Kibish 1 from the Omo River Valley of southwestern Ethiopia, just east of the border with South Sudan and just north of Kenya. It was recovered by a Kenyan team led by the late Richard Leakey (son of Mary and Louis) in 1967. The skeleton was found near the top of Member 1 of the Kibish Formation at a site known as Kamoya's Hominid Site (KHS), named for its discoverer, the famed Kenyan fossil hunter Kamoya Kimeu. It was recently dated to about 233,000 years ago, about 70,000 years later than Jebel Irhoud and possibly contemporary with Florisbad and Ngaloba.[11]

Beginning in 1999, a team led by the paleoanthropologists John Fleagle, Zelalem Assefa, John Shea, and Frank Brown returned to the Omo River Valley to relocate KHS. They were armed with photos from Leakey's 1967 field season, and they not only found the site but recovered additional pieces of bone that fit onto the Omo 1 skeleton, enabling them to place the skeleton in its appropriate stratigraphic context.

If one looks at the Omo 1 skull, which was carefully reconstructed in the 1970s and 1980s by the late British paleoanthropologist Michael Day (1927–2018), one sees that much of it is missing (figure 4.3). In some circumstances, this would bother me, but in the case of Omo 1, the bones that *are* present are highly diagnostic. It has a forward-projecting chin, for example. Its brow ridges are modestly sized; it has a canine fossa, and its occiput is rounded. Its overall cranial anatomy is therefore strikingly modern. A second cranium, Omo 2, was a surface find from about 2.5 kilometers (1.6 miles) to the north, and it lacks a face. Its cranial capacity of about 1,435 cubic centimeters is well above the recent human mean, but its braincase is decidedly less modern, with a long, low, skull and unrounded occiput.

It is *possible* that Omo 2 eroded out of older deposits than the ones in which Omo 1 was recovered, but in the absence of any firm data to verify that supposition, Omo 1 and Omo 2 are best considered penecontemporary. Day and Stringer suggest Omo 2 is a different species from Omo 1.[12] Although this is a possibility, if one finds two fossils at the same location from the same time horizon, I argue that the bar for stating they are members of different species is set much higher. Ultimately, the take-home lesson from Omo (and Jebel Irhoud as well) is that there is sizable morphological variation in these earliest *H. sapiens* populations, with individuals showing different constellations of ancestral versus derived characters.

Our next early *H. sapiens* come from the Afar Region of northeast Ethiopia. First, an adult cranium (BOU-VP-16/1) was discovered by David DeGusta in 1997 at Bouri Vertebrate Paleontology Locality 16. A juvenile cranium (BOU-VP-16/5), about six or seven years old at time of death, was later recovered at the same site. Both crania lay between two tuffs dated by Ar-Ar. According to Clark et al., geological evidence suggests that the crania and associated materials are somewhere between 154 ± 7 and 160 ± 2 thousand years old, although Vidal et al. suggest that they may be more recent still.[13]

As with most early *H. sapiens* specimens, the adult Herto cranium is not completely modern (figure 4.3). It has larger brow ridges than anyone alive today, albeit of the modern discontinuous variety, thick cranial vault bones, and a noncurved occiput. Nonetheless, it also has a reduced facial skeleton that resides underneath the frontal lobes, a more vertical frontal, and a clear canine fossa. Its high, domed cranium has a cranial capacity of about 1,450 cubic centimeters, which is above the recent human mean. Tim White and colleagues named this specimen the holotype of a new human subspecies, *H. sapiens idaltu*.[14] The juvenile cranium, which preserves much of the cranial vault and upper face, became a paratype of this new taxon.[15]

Our last African early *H. sapiens* site has played a central role in the debate surrounding modern human origins. It comes from the mouth of the Klasies River on the Indian Ocean coast of South Africa. First excavated in the 1980s, the site yielded multiple hominin remains from Middle Stone Age levels. In the 1990s, ESR dates for these levels suggested that they all date from about 60,000 to 120,000 years ago, with most dating to 100,000 years ago. The remains are fragmentary, but many show modern affinities: some mandibles have chins, some frontals have brows evincing lateral reduction. Like Jebel Irhoud and Omo, however, specimens within the sample are variable, with some mandibles showing no chin development at all, and one cheekbone (zygomatic) is quite archaic. Years ago I contributed to a study of the ulna from Klasies River, and my colleagues and I concluded it was archaic in its morphology as well.

THE EARLIEST MODERN HUMANS OUTSIDE OF AFRICA

The earliest modern humans to appear outside of Africa are found in the southern Levant, not too far (about 350 kilometers [250 miles]) from humanity's home continent. In 2018, Israel Hershkovitz and colleagues announced they had found the earliest modern human outside of Africa at Misliya Cave. Located on the slopes of Mount Carmel 2.1 kilometers (1.3 miles) from the Mediterranean coast, Misliya Cave is just south of Haifa. The specimen in question, Misliya-1, comes from a Mousterian level. It is an adult left maxilla preserving the crowns of all its teeth except the central incisor. Hershkovitz et al. reported that the maxilla and teeth show that the specimen's affinities lie among other early modern humans and not with archaic forms of the genus *Homo*.[16] A combination of U-series, ESR, and TL dates (some directly on the fossil itself) suggests the specimen is between 177,000 and 194,000 years old. To my eyes, and admittedly I've only seen photos, it looks anatomically modern, but it is a single bone. I hope more fossil material will be forthcoming from Misliya Cave!

The best known early modern humans from the southern Levant were found at Skhūl and Qafzeh. I have already introduced Qafzeh. Skhūl is one of many sites in the Valley of the Caves (Wadi el-Mughara in Arabic or Nahal Me'rot in Hebrew) that the archaeologist Dorothy Garrod (1892–1968) excavated in the British protectorate of Palestine from 1929 to 1934. It lies about 7.5 kilometers (4.7 miles) south of Misliya Cave on the slopes of Mount Carmel, at about 44 meters (144 feet) above sea level. Garrod was educated at Cambridge and Oxford, and in the early 1920s she studied prehistory under the Abbé Breuil in Paris. In 1929, she was picked to direct excavations in the Valley of the Caves, an extraordinary achievement for a woman at that time. In 1939, she became the first woman to win a professorship at the University of Cambridge.

The site's full name is Mugharet es-Skhūl, Arabic for "Cave of the Kids" (baby goats). It yielded three juvenile and seven adult skeletons. ESR dates on bovid enamel suggest ages of about 81 ± 15 and 101 ± 12 ka, and the mean TL date was 119 ± 18 ka.[17] As we saw with the early *Homo sapiens* sites in Africa, there is anatomical variability at Skhūl. The adults with the best-preserved crania are Skhūl 4, 5, and 9, all of whom have large and robust faces. Skhūl 5 is the most studied of the specimens because it has a complete mandible with a slight chin; unfortunately, the middle of the specimen's face is missing. It has a true supraorbital torus, albeit one that is thinner laterally than medially, along with a high, more globular, cranium and rounded occiput. Skhūl 4 has a high forehead and globular cranium but lacks a canine fossa. Skhūl 4, 5, and 6 all have retromolar spaces, a rare occurrence in non-Neandertals, although Skhūl 7 lacks one. Skhūl 5 is the only specimen from the site complete enough to be included in the 3D morphometric analyses of Hublin et al.; it falls within the recent human range of variation for both neurocranial and facial shape.[18]

Qafzeh Cave is farther inland in Galilee, about 31 kilometers (19.5 miles) from the coast and some 220 meters (722 feet) above sea level. It lies within a small hill on the outskirts of Nazareth. It has both Upper and Middle Paleolithic components; twenty-seven human fossils were recovered from its Middle Paleolithic levels alone. The site was first excavated by Moshe Stekelis (1898–1967) and René Neuville (1899–1952) in the 1930s. Excavations recommenced there in the 1960s and 1970s under the direction of Ofer Bar-Yosef and Bernard Vandermeersch. In general, the overall gestalt of the Qafzeh specimens is modern, with variation. The most complete adult crania (Qafzeh 6 and 9) are globular, with rounded occiputs and canine fossae, and according to Hublin et al., both fall within the recent human range of variation.[19] While Qafzeh 6 has a bigger brow than the other Qafzeh specimens, the brow ridges of Qafzeh 3, 6, and 9 all show more modern, discontinuous morphology. In addition, with regard to chins, Qafzeh 8, 9, and 11 all possess the inverted "T" configuration that characterizes most people today.

Thus, from around 300,000 to 100,000 years ago in Africa, and just outside of Africa in the southern Levant, we have fossils that may represent the earliest members of our species, *Homo sapiens*. Some of these specimens, particularly the oldest ones, are characterized by more ancestral (plesiomorphic) characters than we see today, and given their geological age, this is not surprising. Others, particularly some of those later in time, like those from Skhūl and Qafzeh, share derived features (apomorphies) with recent humans; and indeed some, like Qafzeh 9, would likely be lost in a recent human sample were they not so heavily fossilized.

There are also hints of early modern humans in Eurasia beyond the Levant. In 2019, the paleoanthropologist Katerina Harvati and colleagues analyzed two crania from the site of Apidima, Greece, dated to between 170 and 210 ka. The older of these, Apidima 1, lacks a face but has a rounded braincase that clusters more closely with modern humans; the younger of the two, Apidima 2, looks more Neandertal-like in its morphology.[20] Farther away in southern China two sites yielded early modern humans. At Fuyan Cave, Liu et al. found forty-seven modern human teeth in levels dating between 80 and 120 ka; and at Zhiren Cave, Cai et al. bracket modern human remains between 106 and 116 ka. Modern humans appear to have spread rapidly across Eurasia about 100,000 years ago![21]

Keep in mind that if you put ten paleoanthropologists in a room you'll get at least twelve different opinions on the taxonomy of any fossil hominin sample. As a result, not all of these specimens are considered to be *Homo sapiens* by all researchers. In the end, most paleoanthropologists agree that modern-looking humans appear first in Africa, which has implications for the modern human origins debate (chapters 5 and 6).

DEFINING *HOMO SAPIENS*

Unlike every other species of animal, extant or extinct, *Homo sapiens* does not have a holotype, or type specimen, tucked away in a museum cabinet somewhere. Instead, Carolus Linnaeus (1707–1778) wrote this for the genus *Homo*: "*Nosce te ipsum*" (Know thyself). I understand why he said this, but as a practicing taxonomist, I find it unhelpful. In 1959, the British botanist William T. Stearn (1911–2001) proposed that Linnaeus himself should be named lectotype for *Homo sapiens* (a lectotype is a stand-in for the holotype if the latter is lost). Aside from the fact that Linnaeus's body is not lost—no one has removed it from his tomb at the Uppsala Cathedral and placed it in a drawer in a Swedish museum—I see no reason not to honor this suggestion. Were we to do so, the International Code of Zoological Nomenclature (ICZN) would define *Homo sapiens* as "the taxon including the lectotype Carolus Linnaeus that is assigned to the rank of species." We can rest assured that everyone alive today, just a little more than 300 years after Linnaeus's birth, is a member of his species. The problem is that this definition cannot be used when we look at *H. sapiens* across deeper time. It would, therefore, be useful to have a quantitative method for circumscribing our species, at least for fossil specimens.

In a 1982 exploration of the significance of Omo 1 and 2, the British paleoanthropologists Michael Day and Chris Stringer attempted to do just that.[22] In and of itself, this is a great idea, but problems arose from their methodology—problems I would hesitate to call unforeseen. Despite these problems, many of the features I have used as indicative of modern anatomy were originally codified by Day and Stringer. The discontinuous brow ridge morphology is one such feature. They also noted that modern human mandibles have a forward-projecting chin, and that the limb bones of *H. sapiens* tend to be slim and have smaller joint surfaces.

For metric criteria, Day and Stringer argued that the *H. sapiens* cranium has a high vault that is relatively short from front to back. How high a vault is high enough? They said the cranium's vertex radius (a measure of its height) should be at least 64 percent as large as the maximum cranial length. How short must the vault be from front to back? They said the cranial index (cranial breadth divided by cranial length × 100), should be 70 or higher to assign a fossil to *H. sapiens*.

Their method also includes three midline angular measurements depicted in figure 4.4—angles originally defined by the bioanthropologist W. W. Howells (1908–2005). All are measured in the median plane—the plane dividing the body into left and right halves. Day and Stringer posited that a specimen must have a parietal angle lower than 138° to be *Homo sapiens*; a lower parietal angle reflects an increase in size of the parietal lobes of the brain. To calculate the parietal angle, one creates a straight-line chord running from the midline point where the two parietal bones meet the frontal (this point is known as bregma)

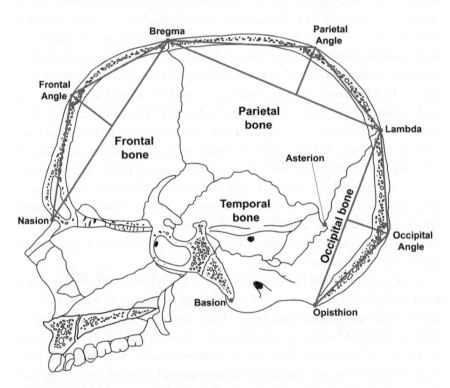

FIGURE 4.4 Mid-sagittal section of a human cranium in left lateral view. Median angles used by Day and Stringer for assigning fossil hominins to *Homo sapiens* are indicated by semicircular arrows. Cranial landmarks discussed in the text are also shown, including the one non-midline landmark on this figure, Asterion, on the right side.

Source: Drawing by author.

to their posterior-most midline meeting point with the occipital (lambda). The maximum deviation of the parietals above this chord (its subtense) is measured, creating a triangle with the chord as its base. The angle at the top of this triangle is the parietal angle. A flatter skull (one with a shorter subtense) will show a higher, more open, parietal angle.

The second angle is the frontal angle. For a fossil to be *H. sapiens*, they argued that its frontal angle must be less than 133°. This angle is calculated in much the same way as the parietal angle. A chord is measured from bregma to nasion; nasion is the midline point at which the frontal meets the top of the two nasal bones. The maximum anterior deviation, or subtense, of the frontal is measured perpendicular to this chord, and a right triangle is created, with the chord as its base. The angle at the subtense's anterior-most point is the frontal angle.

The third angle (occipital angle) is at the back of the skull. Day and Stringer said this angle needs to be greater than 114° for a fossil specimen to be considered *Homo sapiens*. Here a chord is measured from lambda to the posterior midline margin of the foramen magnum, the hole through which the spinal cord passes. This point is called opisthion. The maximum posterior deviation, or subtense, of the occipital bone at midline is measured from this chord and, as with the other angles, serves as the top of a triangle, with the chord as its base. Note that an elongated occipital (as we see in Neandertals) will show a smaller angle than the shorter, more rounded occiput in *H. sapiens*.

To calculate Day and Stringer's final metric, a chord is measured from bregma to points on the lower sides of the cranium close to its back end, at which the temporal, parietal, and occipital bones all meet (asterion). The ratio between the length of this chord and the length of the distance between the right and left asterion (biasterionic breadth) must be greater than 1.19 for a specimen to be considered *Homo sapiens*.

Day and Stringer were careful to point out that not all these criteria will be present in all modern human specimens, but a majority of these characters should be present. Nonetheless, in 1986, the paleoanthropologist Milford Wolpoff questioned the utility of their methodology. He took reported means and standard deviations from Howells's data set to estimate the number of individuals from two Australian aboriginal samples (South Australia and Tasmania) who would be expected to have indices excluding them from *H. sapiens* using the Day and Stringer criteria.[23] Assuming the angles show a normal distribution, Wolpoff estimated that 43 percent of the females and 40 percent of the males from the South Australian sample, and 22 percent of the females and 34 percent of the males in the Tasmanian sample, would have an occipital angle greater than the 114° necessary for them to be assigned to *H. sapiens*. The estimated percent of Australians whose frontal and parietal angles fell outside of Day and Stringer's metric criteria for inclusion in *H. sapiens* was considerably lower but was not zero.

Day and Stringer countered that their method was for fossils and was never intended to be used on recent humans. Also, Wolpoff only had access to tabular data, not individual data for each cranium, so he could not demonstrate that any of the Australian aborigines in the Howells sample had fewer than half of the modern features, as Day and Stringer had stipulated. Wolpoff did, however, have access to fossil skulls, and Late Pleistocene fossil hominin crania from two Australian sites (Coobool Crossing and Kow Swamp) were excluded from modern humans using Day and Stringer's methodology. Despite problems with Wolpoff's analyses, and despite Day and Stringer's best intentions, the damage was done. Any method of circumscribing *H. sapiens* that tends to exclude from our species sizable proportions of groups of living people is clearly invalid. This is particularly problematic if the groups in question have long suffered discrimination and oppression.

A CLASSIC METRIC METHODOLOGY

Could one establish a quantitative method for circumscribing our species that avoids the pitfalls of Day and Stringer? In what I consider a classic 1992 work, the bioanthropologists James Kidder, Richard Jantz, and Fred Smith did just that.[24] First, they established a recent human anatomical baseline. For this, they chose measurements taken by Howells on a geographically variable sample of 595 recent human skulls. Howells took up to eighty-two measurements on each skull, but it is unlikely that all of these will be preserved on fossils. Kidder and colleagues therefore used a subset of seventeen measurements that were more likely to be present in fossils. Even with this reduced set of measurements, however, few fossils have all of them, so they created seven smaller subsets of measurements that enabled them to evaluate increasingly larger samples of fossils. They not only compared the recent humans to likely or potentially early modern humans (such as Cro-Magnon 1, Jebel Irhoud 1, and Qafzeh 9) but also included clearly archaic hominins such as Neandertals and *Homo erectus* specimens as controls.

Quantitatively, they then calculated "log size-and-shape" and "log shape" variables for each set of the measurements they used, following Darroch and Mosimann.[25] Log size-and-shape variables are simply logarithmically transformed measurements. For each individual specimen, log shape variables are computed from these "size-and-shape" measures by creating a ratio between each log-transformed measurement and the geometric mean of all the measurements taken on that individual (the geometric mean is the nth root of the product of n measurements). Darroch and Mosimann argued that the geometric mean of an individual's measurements is the most appropriate measure of that individual's size. Therefore, by creating a ratio between each measurement with the geometric mean for that individual, an isometric size component is removed, creating a "scale-free" variable primarily driven by shape.[26] Size standardization is critical when analyzing fossils because Pleistocene hominins had bigger crania than those of people today. As a single example, would we want to exclude Cro-Magnon 1 from *H. sapiens* solely because he had a bigger head?

Kidder and colleagues then computed distance statistics (Mahalanobis D^2 statistics) on the sample for both the log size-and-shape and log shape data (performing sixteen analyses in total). They were most interested in the D^2 statistic between each fossil specimen and the recent human grand mean. In the end, they considered a fossil to be anatomically modern if it fell within the 95 percent concentration contour of the recent human sample and nonmodern (archaic) if it fell outside of it.

What did they find? First, the La Chapelle-aux-Saints 1 and La Ferrassie 1 Neandertals were used in all sixteen analyses (eight size-and-shape; eight shape—there are equal numbers of size-and-shape and shape analyses for every specimen). They fall outside the recent human 95 percent concentration contour 100 percent of the time. Spy 1 is a less complete Neandertal cranium that was only used in eight analyses, but it, too, falls beyond the recent human 95 percent concentration contour for all of them.

Where do the earliest potentially *H. sapiens* specimens fall? Jebel Irhoud 1 is used in four analyses, falling outside the recent human 95 percent concentration contour for all of them. Skhūl 4 is included in ten analyses; Skhūl 5 in sixteen. For every analysis, these two specimens fall outside the recent human 95 percent concentration contour. Qafzeh 6 could only be used in two analyses. For its size-and-shape analysis, it falls beyond the recent human 95 percent concentration contour, but it falls within it for the shape analysis. Qafzeh 9 was used in four analyses, and it was within the recent human confidence contour for all of them.

What about the Cro-Magnons? For all eight analyses that include size, Cro-Magnon 1 falls outside the recent human concentration contour, and for the shape analyses, it falls outside the modern human range of variation in four of eight analyses. Cro-Magnon 2 was used in four analyses; it falls among recent humans for all of them. Cro-Magnon 3 falls outside of recent humans for all four size-and-shape analyses as well as for three out of the four shape analyses. Dolní

Věstonice 3 was used in sixteen analyses; for only two of these (one size and one shape) did it fall outside the recent human concentration contour.

You have yet to meet the Cro-Magnons from the Czech site of Mladeč; they are about 35,000 years old. Mladeč 1 was used in all sixteen analyses and falls outside of the recent human sample for all of them. Mladeč 2 was used in eight analyses and falls outside the recent human sample for seven of the eight. Finally, Mladeč 5 was used in six analyses and falls outside the recent human range for all of them.

Thus, Kidder and colleagues found that not all potential early representatives of our species consistently fall within the range of variation typical of people today. This is perhaps unsurprising due to the fossils' age and because recent human samples, although geographically diverse, are drawn from a narrow segment of time. Now that you have met the earliest members of our species, in the next two chapters I investigate the evolutionary processes that brought them about.

5

A Paleontological Perspective on Modern Human Origins

"**S**o what would you say is the overall significance, then, of the Lagar Velho child?," the reporter asks my former PhD advisor Erik Trinkaus. It's April 15, 2000, and Erik has just given a talk at the American Association of Physical Anthropologists (AAPA)[1] meetings in San Antonio, reporting new analyses on the recent Lagar Velho discovery. The three of us (Erik, the reporter, and I) are in a busy restaurant along the city's storied and verdant River Walk.

You need to understand that paleoanthropology was in a very different place twenty-plus years ago. Today most of my Anthropology 1010 students come into class already knowing Neandertals contributed genes to modern humans. They know that many of us count a Neandertal among our ancestors, and thanks to readily available genetic tests, some even know their own percentage of Neandertal genes![2] In the year 2000, however, much of the field, and the popular press, held fast to the notion that Neandertals had nothing to do with later human evolution—Marcellin Boule would have been proud! So in December 1998, when two amateur archaeologists, Pedro Souto and João Mauricio, pulled the red ochre–stained forearm bones of a four-year-old child out of a rabbit hole in the Lagar Velho rock shelter 120 kilometers (75 miles) north of Lisbon (figure 5.1), they had no idea what they were unleashing on the field.

The Lagar Velho skeleton was removed en bloc under unpleasant wintry conditions by the bioanthropologist Cidália Duarte and brought back to the lab in Lisbon. Erik was asked by João Zilhão, then director of the Instituto Português de Arqueologia (Portuguese Institute of Archaeology), to study the skeleton. Cidália, Erik, João, and the archaeologists Paul Pettitt and Hans van der Plicht published a paper in June 1999 in which they argued that the Upper

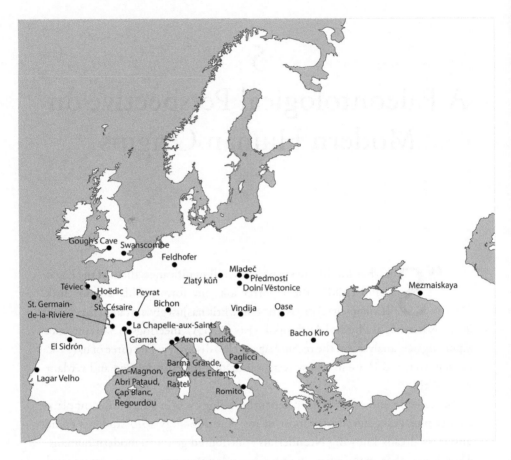

FIGURE 5.1 European and West Asian archaeological and paleontological sites mentioned in chapters 5 and 6.

———

Source: Map by author.

Paleolithic–associated skeleton (uncalibrated radiocarbon date of about 24.5 ka BP; about 30 ka calibrated BP) showed anatomical features indicative of Neandertal ancestry.

Have I mentioned that scientists are a skeptical lot?

The reception of the paper was, euphemistically speaking, less enthusiastic than Erik would have hoped. Now, several months later, after many of us have had the chance to slowly extract the fossil from its matrix and study it in finer detail, we are reporting our findings at the AAPA meetings; that's why the reporter is interviewing us.

Erik explains that he is not arguing that the Lagar Velho child is an F_1 hybrid (one parent a Neandertal and the other a modern human)—the child lived thousands of years after the last Neandertal disappeared from Iberia (chapter 8). Rather, he sees Lagar Velho as part of a modern human population that had earlier experienced interbreeding with Neandertals.

"What this means," Erik says, "is that when these two populations came into contact with each other the differences between them were not..." (he pauses, as if searching for *le mot juste*). It's all the time I need to interject—"Insurmountable."

THE OLDEST QUESTION

After the publication of *On the Origin of Species* in 1859, the year most scientists accepted prehistoric humans as fact, the oldest question in paleoanthropology was born: Where, evolutionarily speaking, did humans like us, our own species, *Homo sapiens*, come from? In 1871 Darwin correctly surmised that our closest living relatives were the African apes, in particular the chimpanzee. This meant that at some point deep in our prehistoric past we and chimpanzees shared a common ancestor. Based on this and drawing from Huxley's 1863 *Evidence as to Man's Place in Nature*, Darwin postulated that if one were to look for fossil remains of our ancestors the place to look would be Africa. Sadly, this idea of Darwin's was ignored for decades as researchers scoured Asia for the so-called missing link between humans and apes. Worse still, many people in Darwin's day ridiculed the notion of a close kinship between apes and us, and some remain uncomfortable with the idea today. In contrast, I side with the Swiss anatomist Édouard Claparède (1832–1871) who, in an 1861 review of Darwin's first book, opined that it was "better to be a perfected ape than a fallen Adam."[3]

By the early twentieth century two opposing camps had arisen in the modern human origins debate, differing primarily in the role they thought Neandertals played in our origins. Aleš Hrdlička (1869–1943), a Czech-American and founder of the AAPA, put forward his "Neandertal Phase" view, in which Central European Neandertals evolved into Cro-Magnons. A similar, but more global model was advocated by Franz Weidenreich (1873–1948), the German-American paleoanthropologist most famous for his work on the Chinese *H. erectus* fossils from Zhoukoudian. In Weidenreich's model, modern humans across the globe evolved from their local predecessors, but all regions are united by gene flow.[4] In his model, too, European Neandertals were probably the primary ancestors of the Cro-Magnons.

In contrast to these views, Boule, his protégé Henri Vallois (1889–1981), and at least for a time, the British paleoanthropologist Sir Arthur Keith (1866–1955) supported what was known as the "Pre-*sapiens*" model.[5] They argued that the *H. sapiens* lineage had great antiquity, evolving first in Asia or Europe, and existing parallel to, but separate from, the Neandertal lineage. Fragmentary fossils from Middle Pleistocene (about 774–129 ka BP) contexts, e.g., a partial cranium from Swanscombe, England, were said to show incipient *H. sapiens* features and an absence of Neandertal traits. These "pre-*sapiens*" fossils and not the Neandertals were the ancestors of the Cro-Magnons. In the 1950s, the American paleoanthropologist F. Clark Howell (1925–2007) "split the difference" between these two camps by arguing that the earlier Levantine Neandertals had undergone a process of "sapienization" and evolved into modern humans, whereas later European Neandertals, the so-called Classic Neandertals, had become geographically and genetically isolated, ultimately becoming extinct.

Beginning in the 1980s, new data, including more fossils and a better understanding of chronology, led to these ideas morphing into today's models: two extreme models of modern human origins (Multiregional Evolution and Recent African Origin) and two intermediate models (Replacement with Admixture and Assimilation). Much of the early scholarship on these models dates to 1984, for in that year a seminal 590-page volume on modern human origins was published, edited by the bioanthropologists Fred Smith and the late Frank Spencer. All four of these models antedate genetic analyses, including those of mitochondrial DNA in living people, which yielded results more consistent with Recent African Origin than with Multiregional Evolution. I mention this because there is a widespread perception, even among scientists, that the Recent African Origin model is rooted in genetic analyses. Although genetic data have transformed the modern human origins debate (chapter 6), the models of modern human origins we use today are *paleontological* models derived from fossil data. Before reviewing them, I'd like to provide some paleontological background on the evolutionary origin of our own genus, *Homo*. For that, we turn to Africa about 2.8 million years ago.

EMERGENCE OF THE GENUS *HOMO*

Homo almost certainly evolves out of the genus *Australopithecus* in Africa around 3 million years ago. It is even possible that the makers of the Lomekwian industry 3.3 million years ago were *Homo*. However, the earliest fossil *Homo* thus far recovered is a 2.8 Ma (Ma = millions of years ago) mandible (LD 350–1) from the site of Ledi-Geraru in Ethiopia's Afar Region.[6] This specimen is cautiously published as "species indeterminant," but it shows affinities to mandibles assigned to *H. habilis*.

As with any fossil taxon, there is disagreement as to which specimens belong in *Homo*, how many species of *Homo* there are, or how these species are circumscribed. In fact, the phylogenetic tree for our genus appears particularly "bushy" between 2.8 and 1.4 million years ago, with some paleoanthropologists recognizing five or more species of *Homo*. In general, *Homo* is said to have larger brains than *Australopithecus*, less projecting faces, smaller back teeth, less powerful chewing muscles, and mandibular molars that are longer than they are wide. The type specimen of *H. habilis* (OH 7) also has a hand skeleton derived for tool use. In this light, there is near consensus among paleoanthropologists that in *Homo* tools are taking over some of the food processing roles that had previously been the job of teeth. Early *Homo* may also show an increase in body size and may have had longer legs and shorter arms than *Australopithecus*, but this is contested.

Some time around 2 million years ago, *H. erectus* evolved, probably in Africa. This species rapidly expanded beyond Africa into Eurasia. We have fossil evidence for *H. erectus* in the Republic of Georgia about 1.75 million years ago and in Java as early as 1.5 million years ago. Although we do not know what species is responsible for their manufacture, we also have stone tools in southern China dated to about 2.1 million years ago (chapter 4). You may come across the name *H. ergaster* in the literature. Some use this taxon to distinguish African from Asian *H. erectus*, but like the paleoanthropologist Susan Antón, I remain unconvinced that these two continental groups are not just regional variants of the same species. Also, the type specimen of *H. ergaster* is a mandible, so we have no clue as to the cranial morphology of that species, thus I avoid using this name.

Homo erectus is a key species in human evolution. It may be the first hominin to expand beyond Africa and with what in geological terms would be considered lightning speed. By 500,000 years ago there are descendants of *H. erectus* across Asia and Europe, including *H. heidelbergensis*, a species almost certainly ancestral to the Neandertals. In fact, if *H. heidelbergensis* lived in Africa too, it could be our ancestor as well. Aside from differences of opinion regarding taxonomic names, none of this is disputed by paleoanthropologists. What *is* disputed in questions of the emergence of *H. sapiens* is what role these descendants of *H. erectus* in Eurasia (whatever we may call them) play in the evolution of modern people.

One final taxonomic point—Wolpoff argues that the emergence of *H. erectus* is the last cladogenetic (splitting speciation) event in our evolution.[7] As such, he sinks *H. erectus* into *H. sapiens* and maintains that evolution in *Homo* subsequent to 2 million years ago represents microevolutionary changes within our own species. I think this creates more problems than it solves, although intellectually I see why it works within Wolpoff's Multiregional Evolution framework.

PALEONTOLOGICAL MODELS OF MODERN HUMAN ORIGINS

The models of modern human origins lie on a continuum with regard to how great a role is played by nonmodern (archaic) humans from outside of Africa. The Recent African Origin model, at least prior to the sequencing of the Neandertal genome, argued that nonmodern Eurasian hominins played no role in the origins of modern humans. In contrast, at the other extreme, Multiregional Evolution argued that the primary direct ancestors of modern humans in any region are the archaic hominins who preceded them there. Two intermediate models (Replacement with Hybridization and Assimilation) fall between these extremes. The intermediate models accept evidence for a recent African origin of modern humans but disagree on how much influence non-African archaic hominins had on our species' emergence.

The Multiregional Evolution model (figure 5.2) is sometimes called the Regional Continuity model because it sees ancestor-descendant anatomical continuity across the archaic/modern human transition throughout the Old World. Its intellectual roots lie in the work of Weidenreich, who emphasized global gene flow as a critical evolutionary mechanism. In its modern form, Multiregional Evolution was first laid out by Wolpoff, Wu, and Thorne in 1984, and Wolpoff has long been its greatest proponent.[8] He views modern human origins not as a speciation event but rather as a series of long-term global microevolutionary processes through which modern human traits become fixed worldwide—i.e., become present in all people everywhere. It is important that these features can emerge at different places and different times but are ultimately globally spread via gene flow. In laying out his model, Wolpoff takes the fact that humans today show geographic differentiation, yet remain members of the same species who freely share genes with each other, and projects it deep into the Pleistocene. Multiregional Evolution is therefore predicated on a balance between gene flow, on one hand, which tends to make populations more alike, and genetic drift[9] and selection on the other, which tend to make populations different. Under Multiregional Evolution, the primary direct ancestors of the Cro-Magnons would be the European Neandertals. However, because it is a *global* process, significant

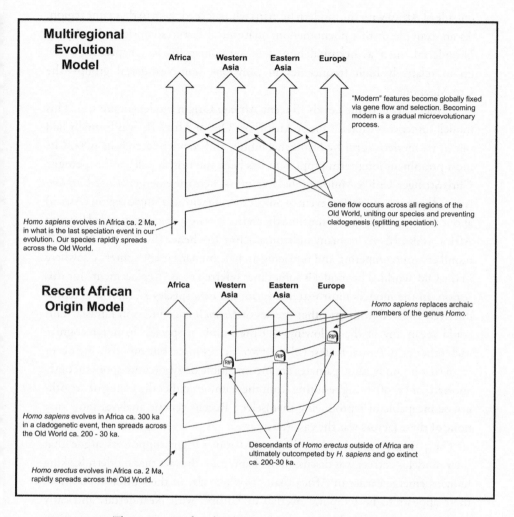

Multiregional Evolution Model

Africa   Western Asia   Eastern Asia   Europe

"Modern" features become globally fixed via gene flow and selection. Becoming modern is a gradual microevolutionary process.

Gene flow occurs across all regions of the Old World, uniting our species and preventing cladogenesis (splitting speciation).

Homo sapiens evolves in Africa ca. 2 Ma, in what is the last speciation event in our evolution. Our species rapidly spreads across the Old World.

Recent African Origin Model

Africa   Western Asia   Eastern Asia   Europe

Homo sapiens replaces archaic members of the genus *Homo*.

Homo sapiens evolves in Africa ca. 300 ka in a cladogenetic event, then spreads across the Old World ca. 200 - 30 ka.

Descendants of *Homo erectus* outside of Africa are ultimately outcompeted by *H. sapiens* and go extinct ca. 200-30 ka.

Homo erectus evolves in Africa ca. 2 Ma, rapidly spreads across the Old World.

FIGURE 5.2   The emergence of modern humans as viewed from the lens of the Multiregional Evolution model (*top*), and the Recent African Origin model (*bottom*).

Source: Drawing by author.

gene flow from other continents into Europe with the evolution of modern people is not only accepted but expected.

Setting aside genetic data for the moment, from a fossil perspective, why would Multiregional Evolution supporters hold that they are correct? Primarily it is because they see evidence for regional continuity outside of Africa. In other words, there are features thought to be genetic that link nonmodern humans outside of Africa with the modern humans who succeed them in that same region. As an example of this phenomenon, anatomical features considered uniquely Neandertal, such as occipital buns and suprainiac fossae (chapter 3), show up in relatively high frequencies in only one non-Neandertal group—the Cro-Magnons.

At the other extreme lies the Recent African Origin model (figure 5.2). This model, sometimes called "Out of Africa" or "Out of Africa II," was formally laid out in its modern form by Stringer, Hublin, and Vandermeersch in 1984.[10] Its most prominent long-term proponent has been the British paleoanthropologist Chris Stringer. Unlike Multiregional Evolution, here the emergence of *H. sapiens* is a cladogenetic speciation event—one occurring within a single region (Africa) about 300,000 years ago. Some time thereafter this new species expanded beyond Africa, as did *H. erectus* many millennia before it—hence the "Out of Africa II" moniker—outcompeting and replacing archaic humans it encounters elsewhere in the Old World. The model is sometimes referred to as "Replacement" for this reason. In the model's most extreme form, the new species *H. sapiens* is reproductively isolated from the other species of *Homo* it encounters in Eurasia. This could mean any of the following: (1) the hominin species do not recognize each other as potential mates; (2) the two species mate, but no offspring occur from that union because gametes are incompatible or zygotes are spontaneously aborted; or (3) offspring resulting from these unions either die young or are otherwise incapable of reproducing themselves. Recent genetic analyses show that none of these factors was the case (chapter 6).

The paleontological reason Recent African Origin supporters argue that their model is correct was discussed in chapter 4—the fact that modern-looking humans emerge earlier in Africa than anywhere else in the world. In addition, there appears to be temporal overlap in Europe, with Neandertals and early modern humans (Cro-Magnons) occupying the continent at the same time (chapter 8). Given the morphological gulf between these two species (chapter 3), suggestions of the latter evolving from the former are deemed impossible.

What of the intermediate models? The German paleoanthropologist Günter Bräuer laid out Replacement with Hybridization in 1984.[11] He agrees with Stringer and colleagues that modern humans first appear in Africa and then expand into Eurasia, but he contends that there they interbred with archaic forms of humans, including Neandertals. It is important to note that he does not view modern human origins as a speciation event; rather, his view is like Wolpoff's in that modern humanity's emergence is due to microevolutionary changes within *H. sapiens*. In terms of genetic contributions to people today, Bräuer posits that the primary ancestry of people today lies with the earliest modern humans in

Africa, with only a sprinkling of genes from archaic humans outside of Africa making it into the modern human gene pool.

The American paleoanthropologist Fred Smith laid out the Assimilation Model in 1984, staking out the middle ground between Bräuer and Wolpoff.[12] In its original 1984 formulation, Assimilation dealt with Central Europe, but it was later expanded into a global model. The primary agent driving modern human origins in Smith's model is demic diffusion of genes from Africa (where modern anatomical features first appear) into other regions. As a key example, Smith argues that modern-looking anatomical features are found in late Neandertals from sites such as Vindija in Croatia or St.-Césaire in France. He says these features are due to the assimilation of extraneous genes into a primarily archaic gene pool. These exogenous genes vary in influence from region to region—in some cases, such as the previous example, the exogenous genes are incorporated into a mostly archaic gene pool, whereas in other regions this extraneous genetic contribution to the local gene pool is likely greater than the local one, or comes closer to Replacement.

BODY PROPORTIONS OF THE CRO-MAGNONS AND MODERN HUMAN ORIGINS

Much of my professional work has focused on the body proportions of fossil hominins. For my doctoral dissertation, I used the body proportions of the Cro-Magnons to test models of modern human origins. Verneau had long ago noted that the distal limb segments (the radius and ulna in the forearm and the tibia and fibula in the leg) of the Balzi Rossi hominins (chapter 1) were elongated.[13] He quantified this via brachial (radius length/humerus length) and crural (tibia length/femur length) indices and found that the Cro-Magnons had high indices—close to those seen in Africans today. In contrast, European Neandertals had low indices, even lower than those of most people in Europe.

In a classic 1981 contribution, Trinkaus demonstrated that among living people brachial and crural indices are significantly correlated with the mean annual temperature of their ancestral regions, such that the lowest indices are found among Arctic populations, and higher indices are found in tropical groups.[14] Armed with these data, he posited that Neandertals were cold-adapted and that the unexpectedly high brachial and crural indices of the Cro-Magnons, despite the glacial conditions under which they lived, were anatomical features suggestive of increased gene flow or population expansion into Europe from warmer climes such as Africa—a pattern more consistent with the Recent African Origin model or an intermediate model than it is with Multiregional Evolution.

Humans are not the only animals to show geographical shifts in limb proportions. Patterns in which limb lengths vary over distance, altitude, or with climatic variables are manifest in many widespread species of warm-blooded animals. This patterning, in which shorter limbs tend to be found in colder climates and longer limbs in hotter climates, was first noted by the biologist Joel Allen in 1877 and is called Allen's Rule.[15] The most common theoretical explanation for this phenomenon is that shorter limbs reduce the body's surface area; because animals lose heat through their skin (the body's surface), shorter limbs help them retain body heat in a cold environment. Conversely, longer limbs increase the body's surface area, thereby facilitating an animal's ability to shed excess body heat in a hot environment.

There are problems with this explanation. Primary among these is that limb proportions seem to have very little effect on human body surface area; bi-iliac (pelvic) breadth has a much greater impact. This was verified by one of my former PhD students, Boryana Kasabova, and me using skeletally derived estimates of body surface area in 2015. Nonetheless, as biologist Ernst Mayr pointed out long ago, ecological rules are merely empirical observations of nature.[16] As such, their veracity does not rest on the validity of any specific theoretical explanation. Thus the fact that multiple bird and mammal species show similar geographical patterns in limb proportions is significant.

Another potential problem arises with using limb proportions as phylogenetic characters: biological evolution is defined as changes in genetic properties of a population across (not within) generations. Therefore, when we argue that changes in anatomical characters reflect evolutionary (read *genetic*) alterations, it is imperative that these characters have a strong genetic component. Just how reflective of the genotype are limb and body proportions? To borrow a tired phrase from social media, "it's complicated." Any morphological feature one can think of is influenced by both genetic and environmental factors—so in that sense, the old question of nature vs. nurture is moot. However, the degree to which environmental factors affect a trait influences whether it should be used in evolutionary analyses. Traits that are too plastic—i.e., those too readily altered during the lifetime of an individual—are avoided. In this case, experimental studies have shown that animals raised in cold environments tend to have shorter limbs as adults than those raised in hot or warm environments.

The anatomist Maria Serrat conducted experiments with mice to see how limb foreshortening comes about in animals raised in the cold.[17] She showed that the foreshortening is due to the pan-mammalian phenomenon in which, upon exposure to cold, blood vessels in the limbs vasoconstrict, reducing blood flow to the extremities, while maintaining blood supply to the vital organs of the trunk, head, and neck. This is a survival mechanism—an effort to sustain high body core temperatures. When vasoconstriction occurs repeatedly during a critical phase in growth and development, the resulting lack of blood flow slows the growth of the limbs at the epiphyseal plates. Serrat's work shows limb proportions can be altered by the environment during growth and development. If they are used as phylogenetic characters, then, it must be done with caution.

Although body and limb proportions may change during the growth and development of an individual, there nonetheless appears to be a strong genetic component to them. A ready example is seen in a measurement taken on living people known as relative sitting height. Sitting height is the measurement of seated people from the top of the surface upon which they're seated to the crown of their heads. Relative sitting height is sitting height divided by stature (standing height). Thus relative sitting height is a good measure of the length of the lower limb in relation to the height of the trunk, neck, and head. Figure 5.3 is modified from Roberts and shows relative sitting height plotted against mean annual

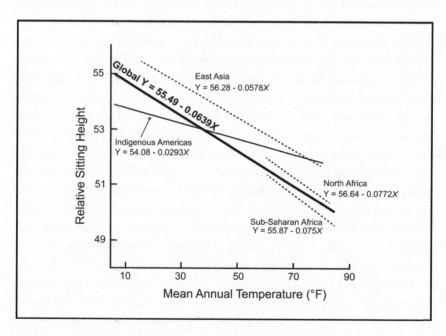

FIGURE 5.3 Variation in relative sitting height (sitting height/stature × 100) among indigenous humans relative to the mean annual temperature of their home regions.

Source: Modified and redrawn by author from Derek F. Roberts, *Climate and Human Variability*, 2nd ed. (Menlo Park, CA: Cummings, 1978).

temperature, contrasting indigenous populations in the Old vs. New World.[18] As evident in this figure, in the Old World, relative sitting height shows a steep negative relationship from cold to hot regions. In colder regions, the height of the trunk, neck, and head on average make up about 55 percent of a person's total height; only 45 percent of their height is due to the length of their lower limbs. In contrast, in hot regions, the height of the trunk, neck, and head is on average about 50 percent of a person's height, with the lower limbs making up the other half. This is variation one would expect following Allen's Rule, with long lower limbs in hotter climates and shorter ones in colder climes.

Now let's look at indigenous Americans—the relationship between relative sitting height and mean annual temperature remains statistically significant, but the slope of the relationship is much shallower. Most Native Americans have short lower limbs, regardless of where they live. This is not due to less environmental variation in the New World; it is because the peopling of the New World involved short-limbed people from Arctic Siberia moving into the Americas from Beringia about 20,000 years ago—a recent event in evolutionary terms. This pattern illustrates two related points: (1) limb proportions in humans are somewhat environmentally stable, at least for multiple millennia; and (2) such broad patterns are not easily erased by changes within the lifetime of an individual.

Might there be a skeletal equivalent of relative sitting height one could use for fossil data? In 1992, Robert Franciscus and I published a study on the *Australopithecus afarensis* specimen "Lucy," for which we estimated the height of her trunk ("skeletal trunk height") from her preserved vertebral elements.[19] I thought I could use the preserved vertebral elements of the Cro-Magnons to examine their limb to trunk proportions to see if they conform to the expected proportions of modern-day tropically adapted groups. Likewise, I could investigate the same relationship in European Neandertals to verify that they show a more cold-adapted morphology. The Cro-Magnons lived in considerably colder periods than today, so if they nonetheless evince heat-adapted limb to trunk proportions, this result would be more consistent with the Recent African Origin model than with Multiregional Evolution, especially if Neandertals were cold-adapted.

I tested this idea in 1997 with a principal components analysis (PCA) of log size-and-shape and log shape variables, as calculated by Kidder and colleagues (chapter 4). PCA is used to find patterning in complex data sets. It involves the calculation of a series of principal components (PC1 = principal component 1; PC2 = principal component 2 . . .). PC1 explains the greatest amount of variation in the data set; PC2 explains the next greatest amount of the variation, and so on. Each principal component is interpreted based on which measurements contribute to it.

In 1997 I showed that the body proportions of four skeletons from Gravettian (box 2.1) contexts, the earliest complete Cro-Magnon skeletons available, lay outside the range of recent Europeans, yet all but one fell within the range of recent sub-Saharan Africans.[20] In contrast, later Cro-Magnons associated with industries such as the Magdalenian either fell in an area of overlap between recent Europeans and sub-Saharan Africans or fell outside of the range of the Africans, but within the European range. I interpreted this as evidence of population expansion or elevated gene flow from Africa with the earliest modern Europeans, followed by subsequent biological adaptation to the glacial cold of Europe.

I recreate that analysis here with a larger sample (table 5.1) and using 95 percent confidence ellipses instead of sample ranges.[21] The measurements analyzed

TABLE 5.1 Recent human sample used in body proportion PCA

EUROPE			SUB-SAHARAN AFRICA		
	M	*F*		*M*	*F*
Bohemia	5	3	East Africa	11	16
Germany	9	8	West Africa	13	3
Norse	4	3	Mbuti*	4	3
Romano-British	14	14	African American	29	28
Medieval France	3	1			
Medieval Bosnia	18	14	**AMERICAS**		
European American	34	33	Point Hope*	48	32
			Puebloan*	2	2
NORTH AFRICA					
Egypt*	6	13	**AUSTRALASIA**		
Kerma*	13	9	Andaman Islands*	1	1
Nubia*	7	5	Australia*	1	0

Sources: Data from Trenton W. Holliday, "Body Size and Proportions in the Late Pleistocene Western Old World and the Origins of Modern Humans" (PhD diss. University of New Mexico, 1995); Trenton W. Holliday and Anthony B. Falsetti, "A New Method for Discriminating African-American from European-American Skeletons Using Postcranial Osteometrics Reflective of Body Shape," *Journal of Forensic Sciences* 44 (1999): 926–30; and Trenton W. Holliday and Charles E. Hilton, "Body Proportions of Circumpolar Peoples as Evidenced from Skeletal Data: Ipiutak and Tigara (Point Hope) versus Kodiak Island Inuit," *American Journal of Physical Anthropology* 29 (2010): 287–302.

*Group included in the PCA but not presented on the scatterplot. There are four Point Hope individuals included for whom sex could not be accurately estimated.

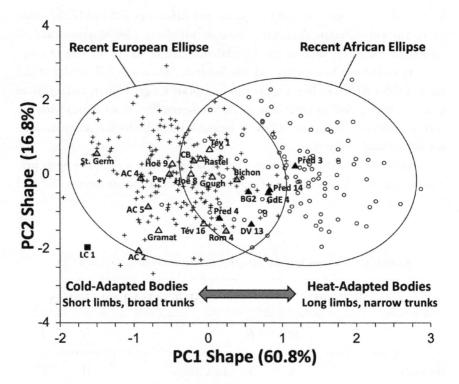

FIGURE 5.4 Scatterplot of PC2 on PC1 (with percentage of variation explained) body proportion "shape" variables. Recent Europeans and European Americans are the crosses; recent sub-Saharan Africans and African Americans are the circles. The La Chapelle-aux-Saints Neandertal is the black square. Cro-Magnons are the black triangles (Early Upper Paleolithic) and gray triangles (Late Upper Paleolithic). Mesolithic people are the white triangles. The full names of the fossil specimens used are found in table 5.2.

are the lengths of the femur, tibia, humerus, and radius (the major limb segments of the body), the diameter of the femoral head (a measure of body mass), skeletal trunk height, and bi-iliac (pelvic) breadth. These features reflect the overall shape of the body. Figure 5.4 is a scatterplot of PC2 (the vertical axis) on PC1 (the horizontal axis). PC1 separates those individuals on the left, who have wider pelves and shorter limbs, from those on the right, who have longer limbs and narrower pelves. Thus PC1 is interpreted as a climatic adaptation component, with cold-adapted people on the left and more heat-adapted people on the right. PC2 does not separate the groups.

Note that recent Europeans and European Americans (the crosses) fall toward the left, and recent sub-Saharan Africans and African Americans (the circles) fall toward the right, albeit with overlap. Among the fossils, the only Neandertal preserving all seven measurements, La Chapelle-aux-Saints 1, represented by the square, has the lowest PC1 score of any of the fossils (i.e., looks the most cold-adapted). Early (>19,000 BP) Cro-Magnon specimens are represented by black triangles. All six have positive PC1 scores, reflecting more heat-adapted bodies, and all lie within the African ellipse (five also fall within the recent European ellipse).

Eight later Upper Paleolithic specimens, dating from about 19,000–12,000 BP, are the gray triangles. Six of them have negative PC1 scores (i.e., are more cold-adapted). Three fall in the area of overlap between the two ellipses, and the remaining five lie within the recent European ellipse and outside the African ellipse. Finally, seven post–Ice Age European Mesolithic specimens, dating from about 10,000 to 6,000 years ago, are indicated by open triangles. Five of them have negative (cold-adapted) PC1 scores. Four lie in the area of overlap between the ellipses, and the other three fall within the recent European ellipse. Thus the earlier Cro-Magnon specimens are more similar to recent Africans in body shape than are the later Cro-Magnons. This result is consistent with an influx of modern people from Africa who over the course of multiple millennia become more cold-adapted in body shape.

Another test can be performed using these same seven measurements. In 1999, I published an article with the forensic anthropologist Anthony Falsetti in which we created discriminant functions for U.S. forensic cases.[22] Our discriminant function is designed to test whether an unknown skeleton is more likely European American or African American. All the females and 87 percent of the males we used to create the discriminant function were correctly assigned to their group (either African American or European American). However, after creating a discriminant function, one needs to test it on a new, independent sample. We did this with skeletons of known ancestry from New Mexico and Florida. Although our function did not perform as well for these skeletons, it nonetheless correctly classified 57 percent of the female skeletons and 82 percent of the male skeletons. Thus despite centuries of significant admixture, European Americans and African Americans still tend to show body proportions more similar to those of their primary continental ancestors.

Table 5.2 shows the results when I ran the fossils through these discriminant functions. Four of the six early Cro-Magnon specimens are assigned to the African American sample with a probability of 83 percent or greater, and Barma Grande 2 and Grotte des Enfants 4 are assigned to the European American sample. Most (six of eight) later Cro-Magnons are assigned to the European American sample with a probability of 76 percent or greater, with two

TABLE 5.2 Discriminant function assignations of fossil hominins and their associated probabilities

Specimen	Sex	Group Assignment	Euro American Probability (percent)	African American Probability (percent)
Neandertal				
La Chapelle-aux-Saints 1	M	European American	99.9	0
Early Cro-Magnons				
Barma Grande 2	M	European American	86.7	13.3
Dolní Věstonice 13	M	African American	16.8	83.2
Grotte des Enfants 4	M	European American	62.2	37.8
Předmostí 3	M	African American	13.8	86.2
Předmostí 4	F	African American	0	99.9
Předmostí 14	M	African American	9	91
Late Cro-Magnons				
Arene Candide 2	M	European American	94.1	5.9
Arene Candide 4	M	European American	99	1
Arene Candide 5	M	European American	76.4	23.6
Bichon	M	African American	18.5	81.5
Cap Blanc 1	F	European American	99.9	0
Peyrat 5	M	European American	92.6	7.4
Romito 4	M	African American	4.8	95.2
St. Germaine la Rivière	F	European American	99.9	0
Mesolithic				
Gough's Cave	M	European American	54.4	45.6
Gramat 1	M	European American	81.5	18.5
Hoëdic 8	F	European American	88.3	11.7
Hoëdic 9	M	European American	92.6	7.4
Rastel	M	European American	74.6	25.4
Téviec 1	F	European American	58.7	41.3
Téviec 16	M	European American	77.3	22.7

Source: Author's own data.

(Bichon and Romito 4) assigned to the African American sample. Finally, all seven post–Ice Age Mesolithic Europeans are assigned to the European American sample. This is a clear temporal trend in which modern humans in Europe show a more heat-adapted body shape and become more cold-adapted over multiple millennia—exactly what one would expect if the Cro-Magnons are derived from an African population moving to colder climes. Whether this reflects population replacement, or that there was gene flow from Neandertals to modern humans during this transition cannot be answered with these data, but genetic data, the subject of the next chapter, provide just such a test.

6

The Genetics of
Modern Human Origins

I t's April 10, 2019, and we're heading north. To our left lies the West
Mesa, a sandy, barren, flat landscape except for the low slopes of its
ancient volcanic cones. I've often thought the mesa could serve as a
location for a movie about colonizing Mars. To our right extends the mile-high
city of Albuquerque, its backdrop the abrupt rise of the Sandia Mountains, pink
and purple in the fading evening light. The Sandias overlook the city from the
lofty height of 3,255 meters (10,678 feet). In the mid-1990s, when a part-time
teaching job at New Mexico Highlands University mandated a twice weekly
commute from Albuquerque to Las Vegas, New Mexico, when I spotted the Sandias just south of Santa Fe on return trips I knew I still had more than an hour
left to drive! Today my former dissertation committee member, Jeffrey Long, a
geneticist and bioanthropologist at the University of New Mexico, is driving me
to Bernalillo. We're meeting Jeff's wife, Colleen Johnson, at Range Café, a restaurant serving traditional New Mexican food. Anytime I'm back in New Mexico,
I gorge myself on the state's culinary delights. Tonight will be no exception.[1]

Jeff, with his round wireframe glasses and bushy gray mustache, is looking
more and more like a nineteenth-century cowboy every year, and he is one of the
most thoughtful people I know. I relish conversations with him because he gets
me to see things in a new light no matter the subject. On this drive we're talking
genetics. Jeff is arguing against the widely disseminated trope that behavioral
differences between males and females are genetic in nature. "The genes in me
have spent half their time in male bodies and half their time in female bodies,"
he says. "If anything," he continues, "behavioral differences between males and
females are an example of the considerable influence sociocultural factors exert
on behavior."

Many of you may be wondering about the influence of the Y chromosome, which spends almost all of its time in male bodies and certainly influences behavior. However, any biologist will tell you that you can't put anything too important on the Y chromosome because only half the population gets one! This segues into the meat of this chapter: as a counterpoint to the Y chromosome, mitochondrial DNA (mtDNA) is almost exclusively maternally inherited because the cytoplasm of the developing zygote is derived from the female gamete, the egg.[2] Many of you already know that mtDNA analysis revolutionized the study of modern human origins. In a 1987 paper on the mtDNA of 147 living people, the geneticist Rebecca Cann and her colleagues transformed paleoanthropology forever.[3] Cann had chosen to analyze her subjects' mtDNA, the specialized DNA found within mitochondria, the organelles known as "powerhouses of the cell." She suspected that mtDNA would be useful for exploring recent human evolution, in part because it lacks the better repair mechanisms present in nuclear DNA (nDNA) and thus evolves at a much faster rate. Also, because mtDNA is almost exclusively maternally inherited, it avoids many of the complicating issues that arise when crossover and exchange of genetic material occur between the paternal and maternal copies of an individual's nDNA during gamete production.

Cann used phylogenetic analysis using parsimony (PAUP), a popular program for phylogenetic analysis, to create the most parsimonious tree for the relationships among these 147 people. The most parsimonious tree is the one with the shortest overall branch lengths; that is, it is the tree with the fewest steps or mutations in it. Parsimony, the idea that one should choose the simplest explanation for any natural phenomenon, dates back to the fourteenth-century scholar William of Occam (1285–1349). Although acknowledging that nature is not *always* parsimonious, parsimony is a good working hypothesis for investigating evolutionary changes.

The tree Cann and colleagues produced had two main branches—an exclusively African branch and one that included both African and non-African people. The branch with the most mutations, and therefore likely the oldest, was the exclusively African branch, so Cann et al. rooted their tree in Africa. Using an estimated mutation rate of 2 to 4 percent per million years, they postulated that the last common ancestor of all surviving human mtDNA lineages lived between 140,000 and 290,000 years ago. By looking at the position of the first branch that contained non-African mtDNA types, they argued for an expansion out of Africa at about 100,000 BP. These results were interpreted as consistent with the Recent African Origin model. No ancient mtDNA lineages that might date back to an earlier expansion of *H. erectus* or *H. neanderthalensis* were found—the mtDNA of all humans everywhere could be traced back only 100,000 to 290,000 years in Africa.

Although undeniably groundbreaking, problems with this analysis were revealed by the zoologist David Maddison in two papers published in 1991.[4] First, using Cann's data, he discovered 10,000 trees that were more parsimonious by five steps than the 1987 "maximum parsimony tree." Many of these trees did not suggest an African root. In the second paper, he pointed out that PAUP uses algorithms for finding maximum parsimony trees that tend to converge on local optima, or islands. Only by chance would these local optima represent *the* global optimum. In fact, although you may find this hard to believe, the order in which data are entered affects the optimum upon which one settles! Thus, to find a global optimum, PAUP must be run many thousands of times with the data entered in a different order each time.

Despite these problems, most genetic analyses in the decades since Cann's study have been consistent with an African-rooted tree. In addition, the structure of human genetic variation is such that there is greater genetic variation today within Africa than elsewhere in the world, and the genetic variation outside of Africa is nested within African lineages. In other words, all non-African *H. sapiens* appear to have arisen from a small number of humans who left Africa sometime in the Middle Pleistocene.

As this well-publicized revolution in the analysis of living human mtDNA was underway, a much quieter revolution was taking place in the study of ancient DNA (aDNA). In the early 1980s, a Swedish graduate student named Svante Pääbo was surreptitiously trying to extract and isolate human DNA from Egyptian mummies. The son of two scientists (his father a Nobel Laureate), Pääbo had a lifelong love of Ancient Egypt but became disillusioned studying Egyptology as an undergraduate (he discovered it was not as romantic as he had hoped). Instead he turned his eye to biomedicine, ultimately pursuing a PhD in immunology. It was in this capacity, as well as through his connections to the Egyptology Department at Uppsala University, that he was able to travel to East Berlin to collect tissue samples from thirty-six Egyptian mummies. In 1985 he published a paper announcing he had extracted, cloned, and sequenced mummy mtDNA.[5] We now know that the DNA he cloned and sequenced was his own. We are fortunate this wasn't known at the time because it might very well have been the end of an extremely productive line of research retrieving, isolating, and analyzing ancient DNA.

In 1997, as a postdoctoral researcher at the University of Munich, Pääbo's team published a paper, the results of which would have been unthinkable even a few years earlier—he and his colleagues extracted and sequenced mtDNA from the original Feldhofer Neandertal specimen.[6] I acknowledge the groundbreaking nature of this paper, but some aspects of it irk me to this day. First, as someone who studies limb bones, the giant section of the right humerus that was removed and destroyed to extract its DNA perturbs me. Second, the pairwise differences between the Feldhofer Neandertal and samples of modern humans and chimpanzees were presented in such a way as to make the Neandertal look almost half as different from people today in his mtDNA as do chimpanzees[7]—a result hearkening back to Boule. In fact, there are two recent humans in the study who are more different from each other than the Feldhofer Neandertal is from at least one of them. In the end, what's more important is what the paper says about interbreeding between Neandertals and modern humans. Although agnostic on the topic of nDNA, the authors argued that mtDNA provides no evidence for interbreeding between the two species.

In 2000 Krings and colleagues managed to extract mtDNA from a late (about 40 ka cal BP) Neandertal from the Croatian site of Vindija. The genetic signature of the Vindija Neandertal was similar to that of Feldhofer. That same year a different team analyzed mtDNA from a Neandertal child's skeleton from the southern Russian site of Mezmaiskaya. It, too, was most similar to Feldhofer.[8]

Recall that Schmitz et al. used historical records to find where the sediments from Feldhofer Cave had been dumped (chapter 3).[9] There they found pieces of the original Feldhofer Neandertal, as well as remains from a subadult and a second adult Neandertal. With Pääbo's help—by this time he was a professor at the Max Planck Institute for Evolutionary Anthropology in Leipzig—they extracted mtDNA from the second adult individual. This, in concert with the data already collected from the Neandertal type specimen, Vindija, and Mezmaiskaya, provided the first opportunity to examine genetic variability among Neandertals. This is important because as you will see in chapter 9, humans today have much lower genetic variability than our closest living relatives—chimpanzees—so the question is whether Neandertals would follow our pattern or be more chimpanzee-like. The new Feldhofer sequence differed for seventeen nucleotides from the closest contemporary human, but there were only one to four nucleotide differences between it and the other Neandertal mtDNA sequences; the average pairwise difference among the four Neandertals was 1.7 percent. Neandertal mtDNA diversity in the Late Pleistocene was therefore more humanlike, and not chimpanzee-like, in character.

Modern human/Neandertal interbreeding was further tested in a 2004 paper out of Pääbo's lab. The first author, David Serre, performed chemical analyses on twenty-four Neandertal and forty Cro-Magnon skeletons.[10] From this, he determined it was worthwhile to try to extract DNA from four Neandertals and five Cro-Magnons. The Neandertals' mtDNA was similar to that of the four other Neandertals already analyzed. In contrast, *none* of the Cro-Magnons had any

Neandertal-like sequences. This would mathematically exclude a large mtDNA contribution to modern humans from the Neandertals, but as previously, it did not address the issue of whether Neandertals had contributed to modern humans' nuclear DNA.

In 2007, the team published mtDNA data extracted from a juvenile humerus recovered from the site of Okladnikov Cave in the Altai Mountain region of southern Siberia.[11] They determined this taxonomically undiagnostic bone had a Neandertal mtDNA signature, expanding the easternmost range of the Neandertals some 2,000 kilometers (1,240 miles) from what had long been the easternmost Neandertal, a juvenile specimen from the Uzbek site of Teshik-Tash. But the biggest surprise of all was yet to come—the retrieval of nDNA from Neandertal specimens! This is surprising because of how unlikely it would be to extract and isolate ancient human nDNA. First, DNA is a fragile molecule—it easily breaks down into small segments, but it survives better in cold than in heat. There is also the problem of contamination. After being buried for thousands of years, upward of 99.9 percent of the DNA recovered from an ancient skeleton is microbial or fungal. Of the tiny fraction of preserved DNA that is human, the vast majority of it would be mtDNA. After all, in most human cells, there is but one nucleus and thousands of mitochondria. Despite these caveats, Pääbo and colleagues dared to try the seemingly impossible, and the way they went about choosing which specimens to analyze is to my mind ingenious.

First, contamination of ancient human DNA with modern human DNA is a genuine concern—recall that Pääbo's first mummy's DNA was actually his own. The dust we see in our homes and offices is largely comprised of human skin cells—most of which retain at least some DNA. To counter this problem, aDNA is analyzed by researchers wearing hazmat or clean room suits in UV-sanitized rooms with positive air pressure. A further complicating issue is that many of these fossil hominins have been handled by dozens, if not hundreds, of researchers and curators over the years, so the chance of one or more of us accidentally "seeding" a fossil with our own DNA is a real concern.

One solution to this problem is the one worked out by excavators at the northern Spanish cave of El Sidrón. They wear clean room suits while excavating to minimize the chances of contaminating any Neandertal remains they happen to uncover. Yet another solution is to scour museums for Neandertal remains that were not identified as such at the time of excavation and were placed among the faunal bones. This is a lot more common than one might think. In 2016 in the Les Eyzies museum, I found a piece of the Regourdou 1 Neandertal pelvis that had been placed in a box marked "*Ours*" (French for "Bear"). The good news for aDNA researchers is that these "faunal" pieces have had far fewer researchers handling them, and therefore should have less modern human DNA contamination.

Once one has selected the Neandertal bones to investigate, the samples are tested to see if sufficient collagen remains. In chapter 4 I mentioned that the primary organic component of bone is the protein collagen, and that during fossilization collagen is replaced by minerals. If there is insufficient collagen (itself a durable organic substance) in a bone, the likelihood that the much more fragile DNA will be present in sufficient amounts to be recovered is vanishingly small. Researchers test the nitrogen content of the bone to determine whether sufficient collagen remains: collagen is a protein, and the building blocks of proteins are nitrogenous amino acids.

If it is determined that sufficient collagen remains, collagen's amino acid chirality is analyzed. In life, amino acids are maintained in their left-handed configuration (chapter 4). After death, these mechanisms no longer function, so the number of right-handed molecules slowly increases over time until an equilibrium is reached between left- and right-handed amino acids. Those who work with ancient DNA have learned that if too high a proportion of the amino acids in collagen extracted from a bone is the right-handed form it is unlikely sufficient DNA for analysis remains.

Finally, before looking for ancient nDNA, a sample of the specimen is used for the extraction of mtDNA. Because we now know the mitochondrial genome of several Neandertal individuals, we can use primers to find specific mtDNA segments. These are then analyzed to see if they are Neandertal-like or modern humanlike. An ideal candidate for Neandertal nDNA analysis is a specimen in which all the above criteria are met that also yields more Neandertal than modern human mtDNA.

In 2006, Pääbo's team published a paper in *Nature* outlining the results of just such an analysis.[12] They used long (119 base pairs) and short (63 base pairs) primers to amplify mtDNA segments extracted from six Neandertal specimens: (1) a late Neandertal from the southwestern French site of St.-Césaire, (2) a Neandertal from Okladnikov Cave from the Altai Mountains in southern Siberia, (3) a specimen from El Sidrón, Spain (the site where archaeologists wear hazmat suits), (4) the juvenile Neandertal from Teshik-Tash, Uzbekistan, and (5 & 6) two femoral shafts from Vindija, Croatia, that had been misidentified as fauna. Of these, four specimens were deemed unlikely to retain Neandertal nDNA. First, 99 percent of the mtDNA from St.-Césaire, both long and short variable segments, was modern. Second, the mtDNA from Okladnikov Cave was 100 percent (long segment) and 98 percent (short segment) modern. Teshik Tash's mtDNA was 100 percent (long segment) and 99 percent (short segment) modern, and finally, 95 percent of the long segment mtDNA of one of the specimens from Vindija (Vi-77) was modern (no short segments were recovered).

In contrast, two specimens were deemed excellent candidates for nDNA extraction. First, although no short segments were retrieved, 75 percent of the long segments extracted from the El Sidrón specimen were Neandertal. The second specimen was the other femur from Vindija (Vi-80)—94 percent of its long segments and 99 percent of its short segments were Neandertal. Two different labs were then given the job of trying to extract nDNA from these individuals.

In 2010, Pääbo's group published a draft sequence of the entire Neandertal genome.[13] I cannot express enough just how astonishing this is to me even more than a decade later. However, some of the group's findings were not terribly surprising. For example, the average genetic divergence between Neandertals and present-day humans, measured as a proportion of the lineage from the human genome to the last common ancestor of chimpanzees and humans, was about 12.7 percent for the autosomes (numbered chromosomes), and between 11.9 and 12.4 percent for the X chromosome. This is about what one would expect based on the presumed timing of the phylogenetic split between humans and chimpanzees about 8 million years ago, versus that of humans vs. Neandertals, about 300,000 to 750,000 years ago.

What was surprising was that Green et al. provided evidence of a Neandertal contribution to people alive today—Asians and Europeans. We would later learn that Africans also retain some Neandertal genes.[14] The proportion of Neandertal ancestry among non-Africans, although higher than that of Africans, was still not great; it was estimated to be between 1.3 and 2.7 percent by one method and between 1 and 4 percent by another. But the fact that some 40,000 years after the Neandertals' disappearance many people today (including me) retain Neandertal genes is amazing.

The year 2010 was full of surprises; Pääbo's team also reported recovering mtDNA from a juvenile fifth distal manual phalanx (the bone at the tip of the pinky) excavated in 2008 at a site called Denisova Cave in the Altai Mountain region of southern Siberia. Denisova Cave is only about 60 kilometers (37 miles) from Okladnikov Cave—the one yielding a juvenile Neandertal humerus. The anatomy of the Denisova finger bone was nondiagnostic. It was expected to be another Neandertal, but analyses of its mtDNA showed that it was about twice as divergent from people today as the Neandertals are! Krause et al. suggested that the Denisova phalanx came from a population that had split from the ancestors of modern humans about a million years ago; in contrast, they argued that the Neandertals split from the ancestors of modern humans around 466,000 years ago.[15] Nuclear DNA sequences extracted from the finger bone and from an adult molar from Denisova are similar to each other, but more important, genes from the Denisova hominins are found today in some people from Asia and Australasia, so these prehistoric people, like Neandertals, are a non-African group whose genes flowed into *H. sapiens*.[16] As there is little anatomy to diagnose a Linnaean species for these remains, they are simply called Denisovans.[17] It is possible they were once widespread across Asia.

Perhaps more surprising, in 2014 Pääbo's team announced that the mtDNA recovered from a toe bone recovered at Denisova was Neandertal.[18] And in 2018, a team reported the discovery of a female Neandertal-Denisova F_1 hybrid they nicknamed "Denny," whose mother was a Neandertal and whose father was a Denisovan.[19] It appears that Denisova represents both the westernmost extent of the Denisovan range and the easternmost extent of the Neandertal range, with these hominins coming into sporadic contact (and mating) with each other.

DO WE FIND EVIDENCE OF NEANDERTAL GENES IN THE CRO-MAGNONS?

Recall that in 2004 David Serre sought to retrieve mtDNA from forty Cro-Magnon skeletons, ultimately succeeding for only five of them: Abri Pataud, Cro-Magnon, La Madeleine, and Mladeč 2 and 25c. In each case, these specimens yielded modern human mtDNA and no Neandertal mtDNA. Similarly, in 2003 David Caramelli and colleagues extracted mtDNA from two about 24 ka Italian Cro-Magnons: Paglicci 12 and 25; no Neandertal mtDNA was recovered. Even as I write these words in 2021, we have yet to find a Cro-Magnon specimen with Neandertal mtDNA. This could mean that the gene flow from Neandertals to early modern humans was male Neandertal with female modern humans because mtDNA is almost always inherited from the mother, or it could be due to lineage extinction—the mtDNA of any mother who has only sons does not survive into the subsequent generation. In any case, we now know that the lack of evidence of an mtDNA contribution from Neandertals does not mean there was no nDNA contribution!

In 2014 in a paper in *Nature*, Qiaomei Fu et al. reported genetic evidence of Neandertal/early modern human interbreeding in a femoral shaft from western Siberia.[20] The femur, which has a modern, non-Neandertal morphology, was found eroding out of about a 45,000-year-old riverine deposit near the town of Ust'-Ishim, some 1,120 kilometers (700 miles) northwest of Denisova Cave. From the specimen's nDNA, they estimated its proportion of Neandertal genes to be 2.3 ± 0.3 percent. This is close to the proportion of Neandertal genes found in living Asians and Europeans (1.7–2.1 percent and 1.6–1.8 percent, respectively). However, they found no evidence for a genetic contribution from the Denisovans. One key finding was that the Neandertal DNA segments of the Ust'-Ishim individual are longer than those of people today. This is because Ust'-Ishim is closer in time to when the admixture took place; therefore, there was less time for the segments to be fragmented by recombination. Assuming there was a single admixture event and a generation length of twenty-nine years, the authors argue that the Ust'-Ishim individual came some 232 to 430 generations after the Neandertal/modern human admixture had occurred. Given the geological age of the specimen, this pins the admixture event to 50,000 to 60,000 years ago, although the authors acknowledge the presence of some longer fragments could indicate a second, more recent, admixture event.

In 2015, Fu and colleagues reported genetic data from the Oase 1 mandible, a 42,000 to 37,000-year-old *H. sapiens* specimen from Romania found deep within a cave known as the Peştera cu Oase, today accessible only by scuba diving.[21] Fu and colleagues found that about 6 to 9 percent of the Oase 1 nuclear genome is Neandertal in origin, more than any modern human yet sequenced. In addition, three of the Neandertal chromosomal segments are so long that the authors believe Oase 1 had a Neandertal ancestor as recently as four to six generations back! To illustrate just how close this is, see figure 6.1, which shows me as an infant with my mother, grandmother, great-grandmother, and great-great grandmother. My great-great grandmother, who died when I was eight years old, is four generations removed from me; it is possible that the Oase 1 individual was similarly removed from his (he was genetically male) Neandertal ancestor.

A year later, the team published the largest analysis of ancient DNA from Europe ever attempted, reporting on DNA extracted from fifty-one ancient skeletons![22] They noted that from around 45,000 to 7,000 years ago the proportion of Neandertal DNA in European modern humans fell from 3–6 percent to around 2 percent; this was interpreted as evidence of selection against Neandertal genes. They also found a high degree of genetic homogeneity among the Cro-Magnons. For example, fourteen individuals dating to about 28,000 to 31,000 years ago from Austria, Belgium, Czechia, and Italy all belong to a single genetic cluster known as the "Věstonice" cluster. This cluster disappeared with the Last Glacial Maximum (chapter 12), an event marked by genetic discontinuity in Europe. The members of the Věstonice cluster are not ancestral to anyone

FIGURE 6.1 Photograph of the author as an infant in 1967 showing five generations of his maternal lineage. The adults from right to left are my mother, my grandmother ("Nana"), my great-grandmother ("Mamasook"), and my great-great grandmother ("Ma"). The Oase 1 modern human may have had a great-great grandparent who was a Neandertal.

Source: Photo by P. W. Holliday; collection of author.

alive in Europe today, but they are related to two genetic clusters that appear in Europe after the Last Glacial Maximum: the "El Mirón" cluster, found in Belgium, Germany, and Spain, and the "Villabruna" cluster, found in Belgium, France, Germany, Hungary, Italy, and Spain. The Villabruna cluster lasted well into the Holocene, including multiple skeletons from the Mesolithic.

In 2021, Pääbo's team published two more papers demonstrating interbreeding between European early modern humans and Neandertals. In the first paper, Mateja Hajdinjak et al. reported nDNA data for three individuals from Bacho Kiro, Bulgaria (chapter 8).[23] Two of these individuals were recently recovered in levels dating to about 45,000 years ago. A third specimen recovered in the 1970s dates to about 38,000 years ago and is thought to be from an Aurignacian layer. In addition to showing evidence for Neandertal admixture back just a few generations, it is fascinating that the two older specimens were more closely related to living people in East Asia and to Native Americans than to present-day Europeans. Indeed, unlike Oase 1 and Ust'-Ishim, these Bacho Kiro individuals appear to have living descendants! In contrast, the later, presumed Aurignacian specimen is more closely related to European Gravettian-age fossil humans (i.e., the Věstonice cluster).

In the second paper, Kay Prüfer et al. reported the recovery of nuclear DNA from the Zlatý kůň female skull found in the Koněprusy cave system in Czechia in 1950.[24] Unfortunately, the specimen cannot be reliably dated because it is heavily contaminated, specifically with organic glue derived from animal sources. To further frustrate matters, Zlatý kůň was associated with early Upper Paleolithic implements that are undiagnostic as to industry! As was the case with Oase 1 and Ust'-Ishim, Zlatý kůň does not appear to have living descendants. Her DNA indicates she has about 3 percent Neandertal genes, comparable to other Upper Paleolithic specimens. However, her Neandertal segments are even longer than those recovered from Ust'-Ishim. This suggests that she had at least one Neandertal ancestor just a few generations back. Based on this, the investigators suspect she may date to the same time as the Ust'-Ishim femur, although this cannot be proven.

In terms of the timing of Neandertal/modern human interbreeding, when one compares the Oase result (interbreeding about 42,000 to 37,000 years ago, likely in Europe) to that of the Ust'-Ishim femur (50,000 to 60,000 years ago, likely in Asia), the Bacho Kiro IUP individuals (about 46,000 years ago, either in Europe or Western Asia), and Zlatý kůň (some time around 40,000 years ago, probably in Europe), it becomes apparent that there were multiple interbreeding events between Neandertals and modern humans in Asia and Europe. However, Oase 1, Zlatý kůň, and the Bacho Kiro IUP individuals fail to share more alleles with later Europeans (including some Cro-Magnons) than they do with recent East Asians. These populations probably did not make a substantial genetic contribution to later Europeans. (It is also possible that any such contribution was "swamped out" with the spread of the Neolithic and agriculture into Europe after 9,000 BP.)

A BRIEF NOTE ABOUT INTERBREEDING AND THE NOTION OF SPECIES

When reading about interbreeding between Neandertals, modern humans, and Denisovans, many of you may be asking: "Doesn't this mean they're all members of the same species?" This idea is due to the long shadow of the eminent biologist Ernst Mayr, who for sixty years argued that species are identified by reproductive isolation. Reproductive isolation means members of different species either do not recognize each other as potential mates or if they mate no offspring result from the union or the offspring that result from such unions die prior to reproductive age or are incapable of reproducing themselves (chapter 5). The classic example of two closely related, yet reproductively isolated, species are horses (*Equus ferus caballus*) and donkeys (*E. asinus*), whose offspring—mules or hinnies—are almost always sterile.

The problem with Mayr's definition of species is that there are many examples of mammalian taxa I would consider to be perfectly "good" species that are differentiated from each other morphologically, genetically, behaviorally, and ecologically but who nonetheless remain interfertile and occasionally, or even frequently, hybridize with each other. Their F_1 hybrids, being rare themselves, tend to mate back into one of the parental populations. Over time and multiple generations, most often their descendants are not morphologically recognizable as hybrids, but their mixed parentage may be revealed through genetic analyses. Examples of such "hidden hybrids" from here in North America include wolves (*Canis lupus*) with coyotes (*C. latrans*), white-tailed deer (*Odocoileus virginianus*) with mule deer (*O. hemionus*), and bobcat (*Lynx rufus*) with Canada lynx (*L. canadensis*). Mayr's definition of species would mandate sinking half of these taxa in the name of species "purity" and ignores important taxonomic variability present in nature.

In chapter 3, I mentioned that Willi Hennig's most important contribution to evolutionary biology is his maxim that only derived characters are evolutionarily informative. In this light, the fact that two taxa who share a recent common ancestor nonetheless remain capable of successfully interbreeding is not surprising—continued interbreeding is merely the retention of an ancestral character in two descendant taxa! I therefore have no problem with species that have leaky boundaries, and my referring Neandertals to *H. neanderthalensis* and Cro-Magnons to *H. sapiens* does not imply that they are reproductively isolated from each other.

WHAT DOES THIS ALL MEAN FOR THE CRO-MAGNONS AND MODERN HUMAN ORIGINS?

The fact that now-extinct groups such as Neandertals and Denisovans contributed genes to modern humans—genes that are still with us tens of millennia later—means one can reject the extreme version of the Recent African Origin model. That said, the fact that modern humans emerge first in Africa long before they are found anywhere else mandates rejection of Multiregional Evolution.

Thus, one of the two intermediate models—either Replacement with Admixture or Assimilation—is correct. Unfortunately, it is difficult to devise a test to reject one versus the other of these intermediate models. The fact that the genetic contribution of Neandertals to people today is so small (4 percent or less) seems more in line with Replacement with Admixture than with Assimilation. Yet the demic diffusion, or elevated gene flow, envisioned by supporters of the Assimilation model may be a more accurate historical description of the initial modern peopling of Europe. As Green et al. so fittingly state, "Obviously, gene flow that left little or no traces in the present-day gene pool is of little or no consequence from a genetic perspective, although it may be of interest from a historical perspective."[25] In the next chapter, I discuss one of the leading ideas as to why modern humans were ultimately evolutionarily successful—that they, unlike their archaic cousins, were capable of "modern human behavior"—a suite of behavioral characteristics giving them an evolutionary advantage.

7

Is There Such a Thing as Modern Human Behavior?

I n August 1993, my sister, her husband, and my parents drove from Louisiana to New Mexico to visit me. In addition to showing them around Albuquerque, I planned a three-day excursion to the northern part of the state. In the Jemez Mountains we visited the idyllic Valles Caldera, a bowl-shaped mountain meadow some 22 kilometers (14 miles) wide, surrounded by evergreen forest—its tranquility belies the catastrophic circumstances of its volcanic birth. We also explored the area around Wheeler Peak, the highest point in New Mexico (4,011 meters [13,161 feet]). While staying in cabins in Red River, a town at the base of Wheeler Peak, an event transpired that affected my view of modern human behavior. After unloading our gear, we decided to take a hike into the ponderosa pine forest. The caretaker's dog, a Chesapeake Bay Retriever, decided he would accompany us. The caretaker told us not to worry; Leon followed hikers all the time. As we entered the woods, Leon hurriedly buried his bone and ran to catch up with us. It wasn't a long hike; we were gone maybe an hour, with Leon leading the way. Upon our arrival back at the cabins, the first thing Leon did was dash over to where his bone was buried, exhume it, and begin to chew on it. It may sound funny to you, but this blew my mind.

I had been in graduate school at the University of New Mexico for five years, and like so many other students, I had fallen under the spell of Lewis Binford (1931–2011). A larger-than-life character, Binford was a big, bearded bear of a man who spoke with a Tidewater Virginia accent. He had participated in the Selma-to-Montgomery marches and was a phenomenal raconteur. Binford was arguably the most important American archaeologist of the twentieth century— he brought scientific and theoretical rigor to the discipline, helping to lead what would become known as "processual archaeology." Lew, as he was familiarly known, could be prickly with his peers, but I assure you that he never "kicked down" and was wonderfully engaging with students. He was convinced that Neandertals, as well as the people who made African Middle Stone Age tools, lacked many of the cognitive capacities we see in modern-day hunter-gatherers, especially in terms of planning and foresight. In particular, he maintained that Neandertals wandered about the landscape without forward thinking, taking game and plant foods they encountered purely by chance. So I had been steeped in a culture in which Neandertals were said to lack the modern capacity to plan ahead, that they didn't or, worse yet, couldn't prepare food to store it long-term, nor were they able to take advantage of seasonal bounties of nuts or salmon runs. Yet here I was watching a *dog* that for all intents and purposes appeared to have used foresight to bury a bone so he could retrieve it again later (Leon became famous in family lore, inspiring my sister and brother-in-law to get their own "Chessie" named Filé).[1]

In chapter 4 I explained the many pitfalls of using anatomy to circumscribe modern humans, especially across deep time. As you might imagine, we encounter similar problems when we try to define something archaeologists refer to as "modern human behavior." Behavior is one of many features, along with anatomical and genetic characters, used by biologists and paleontologists to distinguish one species from another, and *Homo sapiens* should be no exception.[2] This goal is the impetus behind the search for behavioral modernity. However, as with our exploration of anatomical modernity, there is no agreed-upon definition of what behavioral modernity is or, as we shall see, if there even is such a thing!

The principal players in the modern human behavior arena are Paleolithic archaeologists because they have the best, and some might argue the only, data for understanding the behavior of humans in the deep past. When it comes to Paleolithic archaeology, no region has been better explored or studied in more detail than Europe. Paleolithic archaeology was born there, and even today the bulk of the world's archaeologists live in Europe or North America. Europe remains a place where it is geographically and logistically easy for these archaeologists to work. I only half jokingly compare the ease of my own archaeological fieldwork experience in Europe with those of my colleagues who work elsewhere. Depending on where they dig, my colleagues sleep in tents miles away from even the most remote settlements, and everything they eat comes out of a can. In some particularly dangerous spots, they even excavate under the protection of guards wielding Kalashnikovs. I will remain forever grateful to Ann Ramenofsky for giving me the chance to do archaeological fieldwork as an undergraduate, but living in a mobile home in central Louisiana while surveying fire ant–infested milo fields, all while baking under the hot sun in 90 percent relative humidity, is about as far from an Indiana Jones experience as one can get!

In contrast, working in Europe during the summer affords me an opportunity to escape the oppressive Louisiana heat, and I sleep indoors. Fieldwork in southwest France includes chocolate croissants in the mornings (known as *chocolatines* in Aquitaine, the region in which I work) and uncorking a nice Bordeaux in the evening, paired with classic French cuisine and a host of local specialty cheeses— some of the most delicious I've ever had. In Portugal my day begins with delightful custard pastries (*pastéis de nata*). Evening meals may include perfectly grilled sardines or delectable (traditionally chicken) sausages known as *alheiras*, often paired with a nice white wine (*vinho branco*), and always, always, always accompanied by olives. Is it any surprise that we know so much more about the prehistory of Europe than that of the other continents?

The problem, however, is that for well over a century, Paleolithic archaeologists would extrapolate what we knew about the European archaeological record to the rest of the Old World. The issue here is, as the Canadian archaeologist April Nowell notes, when it comes to prehistory, Europe is not a "finishing school" but rather a peripheral cul-de-sac![3]

As with all debates we encounter in prehistory, some background into the intellectual history of these ideas is helpful. In mid-nineteenth century Europe, the Neolithic was defined by technological features such as smoothed/polished stone axes and pottery. In contrast, the Paleolithic was not characterized by its tools, per se, but rather by the large mammals associated with it (recall Édouard Lartet's earlier Mammoth Age vs. later Reindeer Age). Archaeologists soon realized, however, that some tools from the Reindeer Age (the Upper Paleolithic) were technologically more sophisticated than those from earlier periods. Could this indicate greater intelligence among these hominins than among their earlier ancestors or cousins? Here, too, the presence of Neandertals in Europe conspired against us because Neandertals used simpler Middle Paleolithic tools and their cultural remains had yet to show evidence for art, whereas the Cro-Magnons created a much more complex and varied Upper Paleolithic tool kit and made numerous examples of mobile and parietal art (chapter 11).

Paleoanthropologists working in Europe therefore surmised that biology and culture were in lockstep with each other. On one hand, we had the less behaviorally complex and artless Neandertals, and on the other, the artistic and more technologically sophisticated Cro-Magnons. The transition from Neandertals to modern humans in Europe was also conveniently coincident with a transition from the Middle Paleolithic to the Upper Paleolithic. This technological shift was abrupt, at least in a geological sense, and it was frequently referred to as a "revolution." Of course, the simplest and most frequently accepted explanation for the sudden appearance of the Upper Paleolithic was that a new, smarter species of hominin (*H. sapiens*) had come into Europe already bearing it. I call this view the Upper Paleolithic Modernity model.

What, specifically, was revolutionary about human behavior in the Upper Paleolithic? In a classic 1982 article, the archaeologist Randall White (inspired in part by Paul Mellars) argued that the following features distinguish the Upper Paleolithic from the Middle Paleolithic in Europe.[4] (1) A greater proportion of Upper Paleolithic stone tools are made on blades than on flakes; (2) there is more of a stylistic, nonfunctional component to Upper Paleolithic stone tools; (3) there is more emphasis on tools made from antler or bone in the Upper Paleolithic, which could mandate (4) a shift in the Upper Paleolithic toward increased hunting of animals that carry antlers throughout the year. Furthermore, (5) the Upper Paleolithic provides the first evidence in Europe of personal ornaments such as beads; (6) the Upper Paleolithic shows the first artifacts made from "exotic" materials (i.e., from sources 50 kilometers or farther from the site in question), and this could indicate both (7) increased long-distance trade in the Upper Paleolithic and (8) more regular social aggregation among Upper Paleolithic groups. Finally, (9) there may have been greater population densities in the Upper Paleolithic.

Problems with this approach were already evident when White's article appeared. First and foremost, in 1979 at the southwestern French site of Saint-Césaire (figure 7.1), Bernard Vandermeersch and colleagues recovered a Neandertal skeleton (St.-Césaire 1) from within an Upper Paleolithic (Châtelperronian) level (more on this in chapter 8). White is rather nonchalant about this then-recent discovery, arguing that St.-Césaire 1 would only be significant with regard to the emergence of modern human behavior if the Cro-Magnons were the direct descendants of Neandertals. Given St.-Césaire 1's late date (at the time thought to be 36,000 BP), White rejected the hypothesis of a direct evolutionary relationship between these two taxa.

Other shortcomings of the Upper Paleolithic Modernity model involve its extrapolation beyond Europe. In the southern Levant, both anatomically modern, or nearly anatomically modern, humans from Skhūl, Qafzeh, and Misliya (chapter 4), and Neandertals (from Amud, Kebara, and Tabūn) are associated with Middle Paleolithic tools. As such, documenting archaeological evidence in the Levant for a behavioral difference between these two species is difficult, although some, most notably John Shea of Stony Brook University, have attempted to do so.

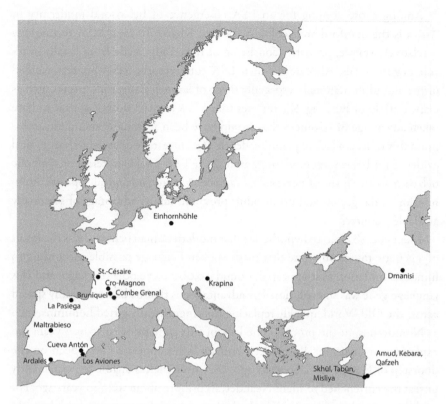

FIGURE 7.1 European and West Asian archaeological and paleontological sites mentioned in this chapter.

Source: Map by author.

Advances in African archaeology also brought to light myriad problems with a blanket application of the European pattern to Africa. Historically, the African Middle Stone Age (MSA), which was characterized by both blades and prepared-core flake tools, was considered contemporary with the European Upper Paleolithic. Starting in the late 1970s, however, new radiometric dating methods showed that the MSA at some African locales dated well over 100,000 years ago, making it contemporary with the European Middle Paleolithic. In spite of these new data, many continued to seek evidence for a modern behavioral revolution in Africa akin to that seen in Europe. Some equated the appearance of the Late Stone Age (LSA) with the emergence of modern human behavior in Africa, essentially lifting the Upper Paleolithic Modernity model and draping it over the African record.

Among those arguing for an LSA emergence of behavioral modernity in Africa is the Stanford archaeologist Richard Klein.[5] To him, MSA technology is relatively simple, primarily consisting of flake tools made from locally available raw materials. Also, unlike their LSA counterparts, MSA hunters avoided prime-age adult mammals, especially those of large or dangerous species such as Cape buffalo or bushpig. Klein notes that MSA hunters do not appear to have taken advantage of resources that would have been seasonally abundant; to his mind this reflects a lack of planning depth. In contrast, he does see archaeological evidence for behavioral modernity among the LSA people in terms of symbolic behavior in the form of personal ornamentation, technological sophistication, hunting of dangerous and prime adult prey, and taking advantage of seasonally available resources.

Klein goes so far as to hypothesize that modern human behavior was the result of a genetic mutation—one that made spoken language possible. According to him, this mutation first appeared around 40,000 to 50,000 years ago, and this language gene was so evolutionarily advantageous that its bearers rapidly spread across the Old World, quickly replacing nonlinguistically gifted hominins, such as Neandertals, in the process. The lack of this gene prior to 50,000 years ago explains why anatomically modern humans, who appear to have been on the doorstep of Europe 100,000 years ago, were nonetheless unable to permanently unseat the entrenched resident Neandertals prior to about 40,000 years ago. For a while, a gene known as FOXP2 looked as if it might support Klein's hypothesis. Humans with a rare mutation in this gene have difficulty speaking, and the human and chimpanzee FOXP2 genes are different. If Neandertals had the chimpanzee version of the gene, it could point to a lack of humanlike language capabilities. Once Neandertal nuclear DNA was recovered, however, it was discovered that Neandertals have the same version of FOXP2 that we have.

H. sapiens is found in Africa and the Levant as early as 300,000 years ago, and nearly modern people are present in Africa by 230,000 years ago (chapter 4). In light of these early dates, Klein argued that "anatomical modernity" must be divorced from "behavioral modernity." He maintains there is no reason to assume that the modern-*looking* fossil hominins from sites such as Omo, Klasies River, Skhūl, or Qafzeh were *behaviorally* modern. In Klein's view, because behavioral modernity is linked to a mutation that leaves no skeletal trace, it cannot be detected from paleontological data alone; it is only evidenced by a shift in the archaeological record.

Some have pointed out that Klein's "mutation" hypothesis would be difficult, if not impossible, to test, and if it cannot be tested, then it is not a scientific hypothesis at all. As was the case with FOXP2, I suspect that it will eventually be tested with genetic data. To my mind, though, there is a bigger problem with Klein's idea. As more and more Paleolithic archaeology was done in Africa, we have come to realize that the record there does not match the predictions of his model. This problem was laid out in the *Journal of Human Evolution* by the archaeologists Sally McBrearty and Alison Brooks in 2000.[6] Their detailed treatise took up an entire issue of the journal and painstakingly laid out the case that modern human behavior did not appear in full form suddenly around 40,000 to 50,000 years ago; instead it appeared incrementally over hundreds of thousands of years.

McBrearty and Brooks argue that the hallmarks of modern human behavior are (1) abstract thinking (the capacity to refer to abstract concepts not limited in time or space), (2) planning depth (the ability to strategize using past experience and then act upon it as a group), (3) behavioral, economic, and technological innovativeness (creation of complex technologies and social networks), and (4) symbolic behavior (representing objects, people, and abstract concepts using arbitrary symbols and reifying those symbols in a cultural context).

How are these modern human behavioral attributes manifest in the archaeological record? They posit that the archaeological signatures of modern human behavior fall into four broad categories: (1) ecological, (2) technological, (3) economy and social organization, and (4) symbolic behavior. In terms of its ecological manifestations, they note that during the MSA we see a range expansion of humans into ecological zones that had been largely or completely uninhabited during the earlier Acheulean. For example, the Aterian MSA industry of northwestern Africa (see figure 2.3) shows human occupation of the Sahara between 90,000 and 40,000 years ago. Similarly, there is an MSA occupation in the tropical forests of West Africa by at least 30,000 years ago, and forests are notoriously difficult places for humans to eke out a living (chapter 12). The MSA also shows an expansion of diet breadth. This means MSA hunter-gatherers are exploiting more species and eating a greater number of foods, ones they had previously ignored or were difficult to catch. This probably included fish; the 90,000-year-old barbed bone points from the MSA site of Katanda in the Republic of the Congo may have been used as harpoons.

In terms of technological evidence for modern human behavior in the African archaeological record, McBrearty and Brooks point out that prismatic blade production is evident early in Africa (about 280,000 years ago) in East Africa's Rift Valley. There is also a short-lived MSA industry in South Africa known as Howiesons Poort that shows evidence of microliths around 70,000 years ago. In addition, MSA bone points from across Africa show that this raw material was frequently utilized. They also point to composite tools—tools made from more than one raw material—and argue that thinning and wear on many MSA points indicate they were hafted onto shafts, greatly increasing their mechanical advantage. In fact, they say hafting was more routine in many MSA contexts at an earlier time than was the case in Europe. Some of these hafted points may have functioned as projectile weapons; they even suggest that small points from the late MSA may have functioned as part of a bow and arrow system—the earliest in the world (chapter 10).

What archaeological evidence is there for economic and social organizational changes with the emergence of modern human behavior in Africa? First, there is long-distance trade or procurement of raw materials in the MSA, with traceable obsidian tools found as far as 190 kilometers (118 miles) from their sources. There is also structured use of domestic space in MSA sites. In addition, intensification of resource extraction took place with increasing technological complexity evident in the MSA across Africa. As for hunting dangerous animals, McBrearty and Brooks agree with Klein that MSA hunters appear to have avoided adult buffalo, which are among the most dangerous herbivores in Africa. But contrary to Klein, they point out that MSA hunters do regularly take down dangerous prey, including zebras and warthogs.

Finally, regarding symbolic behavior, in contrast to Klein, McBrearty and Brooks point out that in the African MSA there is archaeological evidence for self-adornment in the form of beads made from ostrich eggshell or mollusk shells, and pigments, including an important 77,000-year-old incised piece of red ochre from Still Bay (a short-lived, yet technologically complex MSA industry) levels in Blombos Cave, South Africa. Similarly, in 2010, the French paleoanthropologist Pierre-Jean Texier and his colleagues reported the presence of abstract linear depictions on eggshell containers from the Howiesons Poort industry at Diepkloof Rock Shelter in South Africa that date to about 60,000 years ago.[7]

Finally, McBrearty and Brooks argue that most modern human behavioral characteristics appear in the African MSA not all at once about 50,000 years ago but gradually in piecemeal fashion over 200,000 years. As McBrearty later writes, "cognitive capacity for modern behavior was present in earliest *H. sapiens* but . . . it took a few hundred thousand years to put together the package that we now recognize as modern behavior."[8] It is for this reason that she and Brooks call the emergence of modern human behavior "the revolution that wasn't."

There have, however, been critiques of their approach. The Italian archaeologist Francesco D'Errico discounts the importance of blade production.[9] He points out that the Australian aboriginal tool kit of the ethnographic present is based almost entirely on flakes. Similarly, he argues that the first inhabitants of the Americas abandoned prismatic blade technology as soon as they expanded below the Arctic. As an aside, in one of the regions in which I work (central Portugal), Upper Paleolithic assemblages are often dominated by tools made on flakes, with few blades evident.

In a similar critique, the Australian archaeologist Iain Davidson points out that the Paleolithic record of Africa and Eurasia is manufactured by multiple hominin species (what he calls "mixed" records).[10] In many cases, we do not know which hominin species is responsible for the cultural remains at any given site.[11] Davidson maintains the only way to test what constitutes modern human behavior is to look at the peopling of Australia, the first place colonized by a single hominin species (*H. sapiens*) around 65,000 years ago. Also, given that Australia was never connected by land to southeastern Asia, even during glacial maxima when sea levels were at their lowest, humans had to cross 80 to 90 kilometers (50 to 56 miles) of open ocean to reach it. From this, Davidson argues that watercraft manufacture and use indicate long-range temporal planning and thus serve as the earliest evidence for language and modern thinking.

These are all excellent points, but to my mind, the most salient critique of McBrearty and Brooks' model was published in 2003 by the archaeologists Christopher Henshilwood and Curtis Marean.[12] They agreed that modern human behavior in Africa is much deeper than 40,000 to 50,000 years, but their main complaint is that McBrearty and Brooks' list of archaeological signals is Eurocentric and was lifted almost entirely from the European Middle to Upper Paleolithic transition.

Henshilwood and Marean also point out taphonomic biases that arise when comparing the European and African archaeological records. First, when looking at the frequency of bone and antler tools, the two continents differ for two reasons. First, antler is superior to bone as a raw material; it is easier to work and less likely to fracture, allowing for the manufacture of complexly shaped implements. The problem is that aside from the Atlas Mountains in the continent's northwestern extreme, antler is unavailable in Africa. Second, unlike Europe, where much of the continent's ancient basement rocks are alkaline (primarily limestone), leading to preservation of bone in the archaeological record, in Africa most basement rocks are acidic and dissolve bone over time. Where there are limestone cave and rock shelter systems in Africa, most notably in the south, we find the best evidence for the manufacture of bone tools. Thus the comparison of bone and antler tool manufacture between Europe and Africa is unwarranted due to availability and preservation biases.

In addition, Henshilwood and Marean argue that expansion of diet breadth and intensification of resource extraction may be the product of population pressure brought on by environmental change, instability, or population growth.

They point out that many of the prey species "ignored" by MSA hunters would yield little caloric return relative to the number of calories invested in capturing and processing them. Such expenditures include time and energy spent manufacturing weapons, energetic search costs and time spent waiting in ambush, and postacquisition processing time (carcass transport, skinning, defleshing, cooking, etc.). This means that most of the animals missing from MSA sites are those that would be labor-intensive to procure (e.g., fish, flying seabirds). We should therefore not expect to see hunter-gatherers exploiting them, unless other more easily caught or more nutritious prey are unavailable due to competition from other predators, climatic shifts, or a forced move into a less productive environment. Thus the driving economic factor behind a technological or prey choice shift in this case is the amount of labor involved—not the hunters' intellect.

In the end, Henshilwood and Marean offer an alternative definition of modern human behavior that is characterized by symbolic thinking, but more than mere symbolic thought. What is unique about modern humans, they say, is that we use symbols external to ourselves to organize our own behavior as well as the behavior of others. This allows for cultural/informational exchange within and between groups and across generations. How might one recognize this ability in the archaeological record? The best archaeological evidence for external symbol storage is the presence of art and personal ornamentation.

I am swayed by their definition of modern human behavior because I think it captures a key aspect of what makes us human—the use of extrasomatic symbols facilitating the sharing of information with others—even others unknown to us, or unseen by us, across space and time. My issue with their argument is that I do not think modern human behavior as they define it is exclusive to *H. sapiens*.[13] The late Neandertals presumed to be associated with Châtelperronian and Uluzzian industries produced beads and pendants from mollusk shells and mammalian teeth (chapter 8). And it's not just later Neandertals who used personal ornamentation—white-tailed eagle talons from the 130,000-year-old Croatian Neandertal site of Krapina show human-made grooves as if to fix the talons to a strand, and there are even remnants of animal-based fibers and natural colorants recovered from one of these grooves, suggesting it was part of a leather or sinew necklace or bracelet.[14] Similarly, there are raptor claws with cut marks on them from Mousterian levels at Combe Grenal in France.[15] Also, in Mousterian contexts in France and Italy, wing bones of multiple bird species show telltale signs of cutting and scraping to facilitate flight feather removal, which is consistent with their use as adornment.[16]

We now know that Neandertals created other forms of art as well; manganese "crayons" are known from seventy Mousterian levels at forty Neandertal sites in Europe.[17] Perforated mollusk shells stained with colorants were recognized from Mousterian levels at two Spanish sites (Cueva de los Aviones and Cueva Antón).[18] Originally radiocarbon dated at about 50,000 cal BP, uranium-series dates at Cueva de los Aviones suggest that the shells are much older still at 115,000 BP![19] Hoffmann and colleagues also report dates in excess of 64,000 years ago for geometric designs, stenciled hands, and a red-painted speleothem at three cave sites in Spain (La Pasiega, Maltrabieso, and Doña Trinidad [Ardales], respectively), and in 2021 Leder et al. reported a 51,000-year-old "Irish elk" (*Megaloceros giganteus*) phalanx incised with five stacked offset chevrons was recovered from Middle Paleolithic levels at Einhornhöhle, Germany.[20] Most of these dates are older than the earliest evidence for *H. sapiens* in Europe, and given their Middle Paleolithic affinities, they are probably the work of Neandertals. More impressive, Neandertals created circular structures by arranging broken stalagmites deep within Bruniquel Cave near Montauban in southwestern France.[21] These structures strike me as having an aesthetic quality. They were located 336 meters (367 yards) from the cave entrance, so artificial lighting was needed for their construction and use. There was evidence of burnt animal bones within the structures, but the biggest surprise was their age—uranium-series dates of stalagmitic growth since the structures were put in place suggest that they were made between 175.2 ± 0.8 ka and 177.1 ± 1.5 ka! The fact that these structures are so deep within the cave suggests to me that this was a ritual site, and I suspect language was a necessary condition for their construction.

For me, the most parsimonious explanation for the presence of art, personal adornment, and perhaps even ritual in both *H. neanderthalensis* and *H. sapiens* is that some form of language and the capacity for creating art were present in the last common ancestor of these two species. This is not an unreasonable assumption; D'Errico notes that in Europe evidence for pigment use dates to the Acheulean.[22] Although the exact date of the phylogenetic split between Neandertals and us is unknown, paleontological and genetic evidence suggest that it falls somewhere between 750,000 and 350,000 years ago, a period firmly within the Acheulean. Note, too, that although the *capacity* for art may be there its actual *expression* may be triggered by environmental or ecological factors. Also keep in mind that many artistic expressions, such as rubbing ochre on the body, body scarification, or tattooing will be archaeologically invisible.

All of this brings up an eternal paleoanthropological dilemma regarding behavior. We understand that we are evolved from animals that lacked our cognitive faculties. However, we do not know at what point in our evolutionary history cognitive faculties equivalent to our own arose. Using technology as a guide is problematic. Today we are much more technologically sophisticated than the

Cro-Magnons, but it is unlikely we are more intelligent than they were. As the archaeologist João Zilhão states, "'human behavior,' a.k.a. 'culture,' is cumulative, and therefore the passage of time, a.k.a. 'history,' is in itself a powerful explanator (through the buildup of social knowledge and population numbers) of differences between human societies separated by tens of thousands of years."[23]

A semantic issue arises here, however, in the sense that if a nonmodern, archaic species, *H. neanderthalensis*, is characterized by "modern" human behavior, then perhaps we should come up with a different name for said behavior. The whole point of designating a suite of behaviors as "modern" was to distinguish the behavioral repertoire of our own species from that of our closest fossil relatives, the Neandertals.

Did behavioral differences exist between Neandertals and modern humans? Almost certainly. But by the same token, there is a behavioral gulf between people today and the inhabitants of the Cro-Magnon rock shelter 32,000 years ago. As Zilhão, with tongue firmly implanted in cheek, points out, the archaeological paradigm in which we find ourselves insists that the Cro-Magnons have more in common with contributors to *Current Anthropology* than they do with penecontemporary Neandertals.[24] Although this is probably true with regard to skeletal anatomy, he maintains that common sense dictates it cannot be the case for behavior!

My problem with the concept of behavioral modernity is purely a semantic one. In contrast, John Shea takes it a step further, arguing that behavioral modernity as a concept is so flawed as to warrant jettisoning it altogether.[25] He maintains that *H. sapiens* differs from our living ape relatives in how much more behaviorally variable we are. He therefore suggests that we replace "behavioral modernity" with "behavioral variability," which he defines as a behavioral quality reflected in statistical properties of archaeological data such as modes, variance, and skew.[26] As Shea admits, his concept is difficult to model because behavioral variability shifts stochastically across time and space, repeatedly expanding and contracting, with general trends visible only in hindsight.

Although many researchers, myself included, treat these terms as synonyms, Shea maintains that "behavioral modernity" is a *qualitative* condition defined via the presence or absence of specific "modern human behaviors." He then articulates three main arguments. First, both behavioral modernity and modern human behavior are Eurocentric archaeological constructs, and as such are an accident of history. He makes this point with such eloquence and wit that I feel it warrants direct quotation:

> If, for example, the first archaeologists had been Polynesians, the important
> hallmarks of behavioral modernity as they conceptualized it might include

ocean-going watercraft, celestial navigation skills, pelagic fishing, hunting marine mammals, horticulture, domesticated pigs and dogs, ceramics, edge-ground stone axes, monumental architecture, and feather cloaks. . . . Our Poly-nesian prehistorians would probably regard carved antler tools, cave art, and prismatic blade production as quaint local phenomena of no obvious evolution-ary significance.[27]

In contrast to Klein, Shea's second argument is that East African MSA sites dating to the Middle Pleistocene (774,000 to 129,000 BP) show a capacity for "behavioral variability" equivalent to those of more recent humans such as the Cro-Magnons, or the 12,000-year-old Levantine Natufian foragers experiment-ing with horticulture. His third and final argument is that archaeologists' quest to find uniquely derived patterns of behavior exclusive to *H. sapiens* is more likely to succeed if we think about behavior in terms of costs and benefits under a vari-ety of ecological circumstances.

To demonstrate how his behavioral variability model might work, Shea turns to Grahame Clark's Paleolithic "modes" model, in which Paleolithic technology is divided into five distinct (and in Clark's view) temporal modes.[28] In Clark's scheme, Mode 1 tools are simple flake and pebble tools, like those seen in the Oldowan. Mode 2 tools are large bifacially worked tools, like Acheulean hand axes. Mode 3 is identified by the use of prepared-core techniques (e.g., the Leval-lois in the Mousterian). Mode 4 is marked by prismatic blade production, and Mode 5 is identified by the production of geometric microliths—small sharp shards of rock typically embedded into shafts. Shea then examines data from ten East African sites ranging in age from 284 ka through the early Holocene, about 6 ka. Four Middle Pleistocene MSA sites show every mode of production except Mode 5. Only two sites, one Late Pleistocene (LSA) site dating to about 13–17 ka and an early Holocene site dating to about 6–7 ka, show all five modes. Shea argues that this shows equivalent behavioral variability among the sites in question.

As a biological scientist, I am almost always in favor of quantification, but I find myself in agreement with the archaeologists Metin Eren and April Nowell who, in commentaries published alongside Shea's article, point out that trying to measure "variability" is just as problematical as coming up with a "behavioral modernity" trait list.[29] Specifically, before beginning any analysis, the archaeolo-gist has to decide what archaeological traits to measure, which is no less subjective than the traditional "trait list" approach. I worry, too, that comparing within-site or within-level variability in sites of differing temporal depths will create fur-ther problems. A site or level representing two hundred years of occupation is expected to show more archaeological variation than a site or a level representing five or six summers of occupation.

In the end, I argue that modern human behavior, defined as the use of external symbols to organize our own behavior and the behavior of others, is present in at least two species, *H. neanderthalensis* and *H. sapiens*, and therefore the *capacity* for it was likely present in their last common ancestor. The *expression* of that capacity is expected to vary with ecological or environmental pressures, as pointed out by Henshilwood, Marean, and Shea. In many cases, the environmental pressure in question may be the presence of other hominins competing for the same resources. We humans are expert resource extractors, and as such, our population size was rapidly growing, especially in Africa, during the Pleistocene. In light of this, many archaeologists maintain that personal adornment may be a way for people who, due to increasing population density, are much more frequently encountering strangers to communicate within-group membership to those whom they have never met.

BRAIN SIZE, BODY SIZE, AND COGNITION

I began this chapter stating that archaeological data are the best for revealing the behavior of prehistoric humans, but bioanthropology has something to contribute here as well. In paleontology, the volume of the braincase (endocranial capacity) is used to estimate brain mass with great accuracy, since we know the average density of the brain. Brain mass has long been used as a proxy for intelligence, at least across species. That said, for cross-species analyses, one cannot simply use absolute brain size as a measure of cognitive capacity. Here is an obvious example: African bush elephants (*Loxodonta africana*), the largest extant land mammals, have brains four times the size of ours—their brains weigh about 5,400 grams (about 12 pounds), whereas our own brains check in at about 1,302 grams (about 3 pounds). Yet as socially intelligent as elephants are, they're surely not more intelligent than we are! We also have to keep in mind that they weigh in at around 4,301 kilograms (about 9,500 pounds) and are more than sixty times heavier than the average human!

To understand cross-species brain size and its relationship to intelligence, we measure brain size relative to body size, or *encephalization*. Examining simple brain-to-body size ratios, however, is just as misleading as looking at absolute brain size. A small nocturnal primate from Indonesia, Horsfield's tarsier (*Tarsius bancanus*), has a brain mass of 2.7 grams (0.1 ounce) and a body mass of

77.6 grams (2.7 ounces). If we divide brain mass by body mass, we get a value of 34.8. If the average human brain mass of 1,302 grams is divided by our average body size of 58.2 kilograms, we get a much lower value of 22.4. Surely tarsiers are not more encephalized than we are in any meaningful sense of the word!

The reason we cannot use a simple brain-to-body mass ratio for cross-species comparisons is because brain size shows an *allometric* relationship with body size. *Allometry* is defined as the study of size and its consequences, or the study of changes in shape associated with changes in size. It may entail comparisons of adults within or across species, in which case it is called "static" allometry, or it may examine changes in shape that occur during growth and development, in which case it is called "developmental," "ontogenetic," or "growth" allometry. "Evolutionary" allometry analyzes allometric change within and between lineages, and it can be either static or ontogenetic in its approach. For all these cases, the simplest bivariate allometric equation is $Y = AX^k$, for which:

Y = Variable or body structure of interest
A = Constant
X = Body mass
k = Allometric coefficient (exponent)

Curvilinear relationships are harder to model mathematically than straight lines, and larger measurements have absolutely greater variation,[30] so this relationship is often log-linearized to:

$$\log(Y) = \log(A) + k(\log)X$$

This makes the allometric coefficient (k) the slope of a straight line fitted to the data. As long as the two measurements examined have the same number of dimensions, e.g., one volumetric measurement vs. another, such as brain mass to body mass,[31] a slope of 1.0 indicates an isometric relationship. "Iso-" means "same," so isometry is a relationship in which the object of interest keeps exact pace with body size. It also means there is geometric, or shape, identity; there is no change in shape as size increases. In that sense, isometry is the *absence* of allometry. Across mammals, lung volume and limb length are two features that tend to show an isometric relationship with increasing body size.

In contrast, positive allometry is when the allometric slope is greater than 1.0. These relationships are rare in nature because exponential phenomena, like pandemics, can easily get out of control. A classic example of a positive allometric relationship is the Pleistocene "Irish elk" (*Megaloceros giganteus*), whose antlers disproportionately grow in size relative to body size increase. This leads to bigger deer being characterized by disproportionately larger antler racks. These racks, in turn, demand more anatomical features, such as large vertebral spines in the neck and associated musculature to facilitate holding up the now heavier head.

Finally, negative allometry is commonly seen in nature. This is when the increase in size of the object of interest is smaller than the body mass increase (i.e., slopes less than 1.0). Brain size in vertebrates is a classic example of a negative allometric relationship. Across vertebrate species, the size of the brain does not keep pace with increasing body size; the resulting expectation is that larger animals will have *relatively* smaller brains. This phenomenon is illustrated in figure 7.2, a scatter plot I generated of mean $\log_{10}$ brain mass regressed on mean $\log_{10}$ body mass for adults of 630 mammalian species, using data from Boddy et al.[32]

Note that the slope of this relationship is 0.746—if mammalian brain mass were keeping pace with increases in body mass, the slope would be 1.0. Primates and cetaceans such as dolphins and porpoises and are among the more

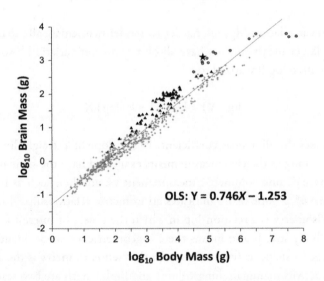

FIGURE 7.2 Log scatter plot of brain mass (g) on body mass (g) for 630 mammalian species. Star is *Homo sapiens*; nonhuman primates are triangles; cetaceans are circles. The ordinary least-squares regression formula is indicated.

Source: Data from A. M. Boddy et al., "Comparative Analysis of Encephalization in Mammals Reveals Relaxed Constraints on Anthropoid Primate and Cetacean Brain Scaling," *Journal of Evolutionary Biology* 25 (2012): 981–94. Plot by author.

encephalized mammals. They are represented by different symbols: black triangles for nonhuman primates and gray circles for cetaceans. You'll notice that primates tend to have larger brains than nonprimate mammals of the same body size, but even among primates humans (the star) are extraordinary in the size of our brain relative to our body, which is reflected in how far above the regression line we lie. Boddy et al. point out that there is *huge* brain size-to-body size disparity among the cetaceans.[33] Dolphins and porpoises tend to fall above the pan-mammalian regression line, whereas the largest of the baleen whales fall well below it.

Related to this observation, growth allometry has revealed an interesting pattern—during mammalian fetal development, brain and body growth tend to show identity; i.e., the brain size to body size allometric slope is 1.0. After birth, however, most mammalian species show slower brain growth, with a shallower, negative allometric slope. This pattern becomes particularly salient in animals whose bodies show marked growth after birth, and it is growth to an enormous body size in the near-absence of brain growth that causes the massive baleen whales to fall so far below the pan-mammalian brain-to-body size regression line. What makes humans unique, even among primates, is that unlike most mammals we continue to show a high rate of brain growth relative to body size growth for several years after birth.[34]

How do we quantify relative brain size in light of the allometric expectation that larger mammals will have relatively smaller brains? The most common way is via the calculation of an encephalization quotient (EQ), which serves as a brain-to-body size ratio, but one calculated within the context of allometric expectations. There are multiple means of calculating EQ. Some scholars produce EQ formulae based on theoretical expectations. For example, Jerison adhered to the idea that brain size scales to body size at the same rate as does the skin's surface area, and he thus expected the brain to show an allometric coefficient of 0.67.[35] The problem is that we see no evidence for a 0.67 brain-to-body size allometric slope in any reliable pan-mammalian data set. For this reason, many researchers instead derive their EQ formulae empirically, using large multispecies data sets. The slopes they observe in their data then determine the allometric coefficient. Boddy et al. do this for their mammalian data set.[36] For the analyses that follow, I adopted their EQ formula:

$$EQ = \text{brain mass in grams} / (0.056 \times \text{body mass in grams}^{0.746})$$

Note that the slope of the line in figure 7.2 is now the allometric coefficient for calculating EQ. Table 7.1 presents EQ values derived from the data in Boddy

TABLE 7.1 Encephalization quotient (EQ) for key
extant mammalian species

Order Primates	EQ*	Rank (Top Ten)
Human (*Homo sapiens*)	5.72	1
Bonobo (*Pan paniscus*)	2.18	
Chimpanzee (*Pan troglodytes*)	1.72	
Western gorilla (*Gorilla gorilla*)	1.31	
Orangutan (*Pongo pygmaeus*)	1.80	
Siamang (*Symphalangus syndactylus*)	2.16	
Celebes macaque (*Macaca nigra*)	3.99	8
Southern pig-tailed macaque (*Macaca nemestrina*)	3.72	9
White-fronted capuchin (*Cebus albifrons*)	4.53	2
White-faced capuchin (*Cebus capucinus*)	4.16	5
Red-faced spider monkey (*Ateles paniscus*)	4.48	3
Horsfield's tarsier (*Tarsius bancanus*)	1.88	
Ring-tailed lemur (*Lemur catta*)	1.29	
Order Cetacea		
Harbor porpoise (*Phocoena phocoena*)	4.43	4
Striped dolphin (*Stenella coeruleoalba*)	4.12	6
Pacific white-sided dolphin (*Lagenorhynchus obliquidens*)	4.10	7
Common dolphin (*Delphinus delphus*)	3.65	10
Bottlenose dolphin (*Tursiops truncatus*)	3.51	
Spinner dolphin (*Stenella longirostris*)	3.38	
Dall's porpoise (*Phocoenoides dalli*)	2.81	
Beluga (*Delphinapterus leucas*)	1.93	
Orca (*Orcinus orca*)	1.49	
Fin whale (*Balaenoptera physalus*)	0.14	
Order Proboscidea		
African bush elephant (*Loxodonta africana*)	1.09	
Asian elephant (*Elephas maximus*)	1.46	

Source: Data from A. M. Boddy et al., "Comparative Analysis of Encephalization in Mammals Reveals Relaxed Constraints on Anthropoid Primate and Cetacean Brain Scaling," *Journal of Evolutionary Biology* 25 (2012): 981–94. Top 10 EQ values are indicated.
*EQ calculated following Boddy et al. using the formula: brain mass [g] / (0.056 × body mass$^{0.746}$ [g]).

et al. Not surprising, humans have far and away the highest EQ (at 5.72) in the entire data set. Six of the top ten mammalian EQ values are from the Order Primates. The second highest EQ (4.53) in the data set is not found among our closest living relatives, the chimpanzees or bonobos, but rather is the white-fronted capuchin monkey from South America! The third most encephalized mammal is also a South American primate, the red-faced spider monkey (4.48). In fact, none of the apes is in the top ten mammals for EQ. In contrast, there are three New World primates and two Old World monkeys, both macaques, in the top ten. Perhaps my taxonomic bias is showing, but I find it surprising that the largest of the "lesser apes," the siamangs, have a higher mean EQ than chimpanzees, gorillas, or orangutans. Indeed, the gorilla EQ (1.31) is only slightly higher than that of the ring-tailed lemur (1.29).[37] Recall that a simple brain mass to body mass ratio made Horsfield's tarsier look more encephalized than humans. This species' EQ (1.88) is much smaller than that of humans but is higher than that of chimpanzees (1.72).

Beyond the primates, cetaceans hold the other four positions in the EQ top ten, with the most encephalized, the harbor porpoise, taking the fourth spot overall at 4.43. You'll find in table 7.1 that many species of dolphins and porpoises have higher EQ values than the apes. In contrast, the giant fin whale has the lowest EQ (0.14) of any of the 630 species in the pan-mammalian data set.

Figure 7.3 is a box-and-whiskers plot of EQ for the 630 mammalian species. The left edge of the box represents the 25th percentile of the sample (0.73). The median, or 50th percentile, is 0.97, and is represented by the vertical line within the box. The right edge of the box is the 75th percentile, which is 1.24. Thus the box spans what is known as the interquartile range, or IQR—the middle 50 percent of the sample. Assuming a normal distribution, the expected maximum and minimum for the samples are represented by the whiskers (the horizontal lines that extend out from the box), the lengths of which are calculated as 1.5 × IQR. The right (high) whisker is 2.005, and the left (low) whisker would be −0.035 but is instead the observed minimum EQ, the fin whale at 0.14. Notice that there are multiple high-EQ outliers, represented by circles, lying beyond the right whisker. This means the mammalian EQ distribution is highly skewed—a strikingly nonnormal distribution. What amazes me, however, is just how far from any other species the human EQ of 5.72 is. The take-home message of these EQ values is that for our size human brains are much, much larger than expected. We humans are, forgive the term, freakishly encephalized animals.

In 1997, Chris Ruff, Erik Trinkaus, and I published a paper for which we estimated brain and body mass for Pleistocene *Homo*.[38] Brain mass was estimated

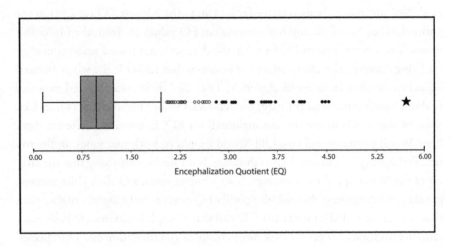

FIGURE 7.3 Box-and-whiskers plot of EQ, 630 mammalian species from Boddy et al., Outliers that lie more than 3 times the interquartile range above the right whisker are depicted as black circles. *Homo sapiens* is the black star to the far right.

Source: Data from A. M. Boddy et al., "Comparative Analysis of Encephalization in Mammals Reveals Relaxed Constraints on Anthropoid Primate and Cetacean Brain Scaling," *Journal of Evolutionary Biology* 25 (2012): 981–94. Plot by author.

from endocranial capacity; body mass was estimated from femoral head size or stature and bi-iliac (pelvic) breadth.[39] Keep in mind that fossils tend to be fragmentary, so only rarely are we able to acquire body mass and endocranial capacity data for the same individual. For this reason, our method involved estimating a mean body mass and a mean brain mass for temporal/taxonomic groups of *Homo*. We compared our brain and body size estimates, and EQ values calculated from them, to recent humans. In the 1997 paper we used Martin's EQ formula;[40] here I recalculated EQ for our sample using the Boddy et al. formula.[41] The relevant summary data are presented in table 7.2. You'll notice that the recent human mean EQ value (6.48) is even higher than the already astronomically high 5.72 EQ reported in Boddy et al.[42] This is because our brain mass estimate is larger than Boddy's (1,302 grams vs. 1,250 grams), and our body mass estimate is lower (65.1 kilograms vs. 58.2 kilograms [143 pounds vs. 128 pounds]).

Note that in table 7.2 all Late Pleistocene hominins—Neandertals, Skhūl-Qafzeh, and Cro-Magnons—have brains that in absolute size are larger on average than those of people today, but in every case, this is also in the context of larger mean body sizes. The EQ data paint an interesting picture. You'll see that

TABLE 7.2 Mean brain mass, body mass, and encephalization quotient (EQ) for recent humans (and estimates for fossil hominins)

GROUP NAMES	Brain mass (g)	Body mass (kg)	EQ*
Recent Humans:			
Global Sample	1,302	58.2	6.48
LATE PLEISTOCENE (129 ka–11.7 ka)			
Cro-Magnons:			
Late Upper Paleolithic (11.7–21 ka)	1,412	62.9	6.64
Early Upper Paleolithic (21–35 ka)	1,460	66.6	6.57
"Proto-Cro-Magnons":			
Skhul/Qafzeh	1,444	66.6	6.50
Neandertals:			
Combined Europe/Western Asia	1,399	72.9	5.89
MIDDLE PLEISTOCENE (774 ka–129 ka)			
Homo heidelbergensis (Europe/Africa)	1,177	68.3	5.20
Homo naledi (South Africa)	512.5	45.9	3.24
Homo floresiensis (Indonesia)	425.7	32.7	3.26
EARLY PLEISTOCENE (2.58 Ma–774 ka)			
African *Homo erectus***	871.0	59.8	4.25
Dmanisi (Georgia) *Homo erectus*	639.0	45.3	3.83

Sources: Data from Christopher B. Ruff, Erik Trinkaus, and Trenton W. Holliday, "Body Mass and Encephalization in Pleistocene *Homo*," *Nature* 387 (1997): 173–76.
*EQ calculated following Boddy et al. using the formula: brain mass [g] / (0.056 × body mass$^{0.746}$ [g]).
**Taxonomy of postcranial remains uncertain. The endocranial capacity data are limited to *H. erectus/ergaster*, but body mass data could include *H. rudolfensis* or *H. habilis*.

the Cro-Magnons have the highest average EQs (6.64 and 6.57), making them the most encephalized hominins of all time, and Neandertals have the lowest EQs among Late Pleistocene hominins. I cannot help but note, however, that the Neandertal EQ estimate (5.89) is higher than the recent human EQ value (5.72) from Boddy et al.![43] Skhūl/Qafzeh (6.50) also show a high EQ, falling between those of Neandertals and Cro-Magnons.

After publication of our 1997 paper, some pointed to the Neandertal EQ value as indicative of a cognitive difference between the Neandertals and us. I urge the reader to avoid such an overinterpretation of these data. Among closely related taxa, we do not expect to see the same functional differences due to relative encephalization, or even the same slopes, one finds across a broad array of mammals. For example, within humans and breeds of dogs, the brain-to-body size allometric slope tends to be much shallower at about 0.22.[44] Also, although the Cro-Magnons have the highest mean EQ among the hominins, I do not believe for a minute that they were smarter than people alive today. By the same token, I do not believe that the Neandertals were less intelligent than we are—imagine people with the global average brain mass of 1,302 grams. Any of them with a body mass of 67 kilograms (148 pounds) or greater have a lower EQ than the Neandertal average!

What about Middle Pleistocene *Homo*? Before discussing the Middle Pleistocene EQ data in table 7.2, I should mention that, following archaeologist Glynn Isaac (1937–1985), many of my colleagues call the Middle Pleistocene the "Muddle in the Middle" because we cannot come to an agreement on how many species of *Homo* there were then, or how to circumscribe them. In table 7.2 I have grouped several Middle Pleistocene fossils from Europe and Africa into the species *H. heidelbergensis*—a problematic taxon for multiple reasons. First, its holotype is a mandible, so we have no idea what its cranium looks like. Second, some maintain *H. heidelbergensis* is limited to Europe; these scientists refer African Middle Pleistocene fossils to other available species such as *H. rhodesiensis* or *H. helmei*, and they may be right in doing so. All the fossils I placed in *H. heidelbergensis*, however, are characterized by brain sizes comfortably within the nonpathological recent human range, and within the context of large and heavily built skulls and bodies. As a result, the *H. heidelbergensis* mean EQ of 5.2 falls just outside the range of people today, but nonetheless it remains far higher than that of any extant mammalian species (except *H. sapiens*).

There are two surprisingly small-brained Middle Pleistocene forms of *Homo* who are so different from their fossil cousins that most paleoanthropologists give them species status: *Homo naledi* and *H. floresiensis*.[45] These two taxa, with EQs of 3.24 and 3.26, respectively, are more encephalized than any of the extant apes but are nonetheless less encephalized than multiple species of monkeys and dolphins; neither would make our pan-mammalian "Top Ten" list!

What about Early Pleistocene *Homo*? Here, too, we have issues. First, taxonomic complications exist with African *H. erectus*. There are few associated skeletons for this species, and multiple hominin species roamed Africa in the Early Pleistocene. This means a 1.8-million-year-old African femur cannot be reliably assigned to a species—it could belong to *H. erectus* (or *H. ergaster* if that is, in fact, a separate taxon), *H. rudolfensis*, or *H. habilis*. There are also australopith species at 1.8 Ma—members of the genus *Paranthropus*. Although their femoral anatomy is distinctive from *Homo* (*Paranthropus* tend to have smaller femoral heads and longer femoral necks than *Homo*), an incomplete or damaged australopith femur could be mistaken for *Homo*. The South African species *Australopithecus sediba* is also present at about 1.98 Ma; we cannot rule out it surviving to 1.8 Ma. Despite these caveats, the EQ I calculate for African *H. erectus* (4.25) is lower than that of two New World monkey species and a species of porpoise (table 7.1). Were they extant today, African *H. erectus* would hold fifth place in mammalian EQ. The final fossil hominins examined here are the 1.75-million-year-old *H. erectus* sample from Dmanisi, Georgia. They tend to be smaller in both brain and body size than African *H. erectus*; their EQ of 3.83 would give them the ninth position in the Boddy et al. data set, between two species of macaques!

What should we make of these fossil EQ data? I would interpret them this way: (1) there are probably no significant cognitive differences between recent humans, "Proto-Cro-Magnons," Cro-Magnons, and Neandertals; (2) in contrast, significant cognitive differences between recent humans and early Pleistocene *H. erectus* are probable; (3) there are probably significant cognitive differences between people today and the small-brained species *H. naledi* and *H. floresiensis*; and (4) it is impossible to tell if cognitive differences would exist between people today and the larger-brained Middle Pleistocene forms of *Homo*, referred here to *H. heidelbergensis* (although I cannot rule out a small cognitive difference).

There is but one final caveat to mention before concluding this discussion on EQ. Brain size and EQ are blunt instruments when it comes to anatomical correlates of intelligence. In this regard, I distinctly remember a dinner conversation thirty-odd years ago in which the paleoanthropologist Bruce Latimer took me to task for arguing that dolphins were more intelligent than apes. He insisted that much of a dolphin's brain is dedicated to the mechanics of sonar interpretation, not to intelligence per se (and he may have a point). What, then, are better anatomical indicators of intelligence? As reviewed in Roth and Dicke,[46] anatomical and physiological factors such as cerebral cortical area,[47] intrinsic organization of the cortex, neuronal density, and information processing capacity (IPC) are all more highly correlated with intelligence than brain size or EQ alone is. These features are unavailable in the fossil record, but ancient DNA may ultimately provide insight into the brain function of our Pleistocene cousins.

In fact, in 2021, a group of researchers revealed experimental results in which they had inserted an ancestral version of the neuro-oncological ventral antigen 1 (NOVA1) gene, one present in Neandertals and Denisovans, into human stem cells.[48] Modern humans have a mutation in NOVA1 that is unique to our species, and because this gene is primarily active only in developing brains, the researchers suspected it might be a candidate for a functional neurological difference between us and Neandertals. In their experiments, these genetically modified stem cells formed organoids, clusters of human neurons raised in vitro. Under magnification, the surface of these modified organoids was different from normal human organoids, but more interesting, they differed in the proportion of neuron types, showed different electrical signaling properties, and began propagating signals at a later date than the normal human control organoids. Thus, there may be physiological/neurological differences between Cro-Magnon and Neandertal brains that are not evident from gross anatomy alone. This research is in its infancy; more results are almost certain to be forthcoming.

A FINAL QUESTION: WHY ARE WE HERE AND NEANDERTALS AREN'T?

I have spilt a lot of ink arguing that Neandertals had the capacity for modern human behavior and cognition, even if that capacity is not always expressed in archaeologically visible ways. This is due in part to the fact that the use of arbitrary extrasomatic symbols to communicate with others is contingent upon ecological, environmental, and demographic factors, such as increased population densities and the frequency with which one is likely to encounter people one does not know. As archaeologists who specialize in the African MSA will tell you, this is just as true for *H. sapiens* as it is for *H. neanderthalensis*, and it may be the reason technologically complex, yet short-lived MSA industries such as Still Bay and Howiesons Poort make brief appearances.

I already imagine what some of my archaeological colleagues reading this chapter are thinking: If modern humans did not have some kind of behavioral or cognitive advantage over the Neandertals, why are we still here and Neandertals aren't? I don't have a definitive answer to that question. It's entirely possible that Neandertals had the same capacity for modern behavior as the Cro-Magnons but were victims of extraordinarily bad luck. We know they faced additional resource competition from the Cro-Magnons at the worst possible time—during the bitterly cold Heinrich H5 event at about 45–40 ka. If Neandertals were already living at low population densities, this may have been the proverbial straw that broke the camel's back. A related hypothesis I favor is that modern humans had a demographic edge over the Neandertals. Perhaps the Cro-Magnons had higher fertility due to shorter interbirth intervals resulting from shorter gestation length than Neandertals—an idea proposed by Trinkaus in the 1980s—or an earlier weaning age. Another possibility is that the Cro-Magnons had slightly higher infant survivorship. Perhaps they were more resistant to certain diseases, or maybe they had slightly longer life spans or reproductive periods. In a now classic work, the archaeologist Ezra Zubrow demonstrated that even slight changes in demographic parameters such as these could see the complete replacement of Neandertals by Cro-Magnons in as few as 1,000 years.[49]

How might such demographic advantages come about? The paleoanthropologist Steven Churchill has put together an elegant argument on how modern humans outcompeted Neandertals.[50] In table 7.2, you'll notice that the Neandertals have the largest mean body mass of any late Pleistocene hominins. Churchill opines that having a large, metabolically expensive body and brain means Neandertals had to spend more time foraging for food to secure sufficient calories to keep their internal furnaces lit. This was made worse by the fact that they lived at high latitudes where Neandertals faced greater exploitation competition from other carnivores because more of the Neandertal diet had to come from animal sources (plant foods at high latitudes are simply unavailable for much of the year). As a result, the carrying capacity of the environments in which they lived was depressed, and aside from certain periods such as interglacials and warmer interstadials (chapter 12) during which the climate ameliorated, Neandertals lived at low densities and were rare on the landscape.

Churchill maintains that the body mass of archaic humans in Africa during the Middle Pleistocene was similarly large, but that exploitation competition from other members of the carnivore guild was lower because in the tropics and subtropics more of the human diet could come from plants. He posits that in the later Middle Pleistocene, with the emergence of *H. sapiens* in Africa, one does see a reduction in body size. This, coupled with their development of projectile weapons (chapter 10), probably led to increased fertility, longevity, and carrying capacity for these earliest *H. sapiens*, the result of which was a rapid population size increase. For Churchill, the expansion of *H. sapiens* beyond Africa appears to have been more demographically than environmentally driven.

In the end, when Neandertals encountered the Cro-Magnons, Churchill says that the former may have adopted some of the cultural innovations of the latter (such as long-range projectiles; chapter 10), but Neandertals were already living at lower population densities and still had to fuel their heavy bodies, so they simply may not have been up to the task of keeping demographic pace with the new arrivals.[51] The next chapter explores the fascinating time in Europe during which Cro-Magnons and Neandertals coexisted.

8

Neandertal and Cro-Magnon Interactions in Europe

The pair were miles from home. It was winter, and siblings Parga (nineteen) and Lillo (sixteen) were hunting in the shadow of the inland mountains. Both were unmarried, but they secretly hoped to find mates this summer at the Feast of Eleven Clans—an enormous gathering of more than four hundred people marked by food, music, games, and contests. It was the best time of the year, but as the chill wind rushed down from the mountains, summer seemed very far away indeed. The two were on the third day of a six-day mission to hunt big game to bring back to the coast where their group was overwintering. Thus far they had killed one red deer, a buck, which they had butchered, smoked, and cached in a hole dug near the small rock shelter that served as their base of operations here in the highlands. They were currently tracking another red deer—its tracks in the snow had been heading east, as if to the mountains, but then it suddenly turned to the north.

Heading north here could be an issue. A few hundred meters to the north was the land of the stout ones their people called Orvs. Neither Parga nor Lillo had ever seen one, but Orvs were said to be shorter than people yet as strong as bears. They were also said to be as pale as an arctic fox, with hair on their heads the color of fire. Parga and Lillo's father, Tol, had warned them to never approach an Orv. Like his children, Tol had never seen one, but his father's mother had when she was young. As Tol's grandmother told him, she had been foraging in a woodland when an injured female Orv appeared seemingly out of nowhere. She gave the Orv some smoked meat and berries and rubbed some healing herbs on the Orv's wounds. She had then gone to find more herbs, but when she returned the Orv was gone. Their father told them that their great-grandmother had been lucky it was a female Orv—male Orvs were strong enough to kill you with their bare hands if you were foolish enough to make them angry.

Now Parga and Lillo were faced with a difficult choice. Turn around and hope to find suitable prey elsewhere or follow this red deer just a tiny bit into Orv territory, kill it, and quickly bring it back to the rock shelter to butcher it. No one would ever know they had ventured into "Orv country." Besides, they couldn't be sure any Orvs were left in the area, or even if Orvs were real—not to mention the fact that a second red deer would be a huge prize to bring back home. However, venturing into Orv territory was forbidden and potentially dangerous. What should they do? Lillo pulled a chamois ankle bone out of a satchel. Three of its sides had been stained black with charcoal, the other three red with ochre. "Here's what we'll do: I'll toss this bone. Black side up, we turn around; red side up, we follow the deer." The die was tossed and landed red side up. The two silently started making their way north.

An hour later, as they were following the tracks up a ridge, a loud CRACK surprised them. They immediately dropped to the ground and crawled on their bellies through the snow like snakes. Slowly they stuck their heads over the ridge and peered into the valley below. There, to their amazement, about 30 meters away, sat five Orvs, a mix of males and females, using stone tools to break open the bones of the red deer Parga and Lillo had been tracking. There was another loud crack—an Orv had just broken open a bone and was removing the marrow with his fingers. All the Orvs wore loose-fitting cloaks of red deer hide. Their skin was indeed pale, although not as white as snow—it was more akin to the color of people's palms—and four of them had hair the color of fire!

Three Orvs had their backs to them, but they could see the faces of the other two. They had huge noses, thick brows, and when they spoke, you could see they had small pegs for front teeth. A child Orv appeared out of the woodland and said something to the adults, which made them all laugh. One of the adults ruffled the child's hair and gave him some marrow. It was a much more tender action than Parga and Lillo had expected to see, but they also didn't want to wait around to see any examples of the Orv's legendary violent side. They shot each other a knowing glance and silently slunk backward down the slope. As they did, loud cracks continued to echo in the valley, so they were confident that the Orvs had not spied them. Once they were certain they were out of earshot, they turned and ran as quickly as they could back to the rock shelter. They arrived out of breath yet strangely exhilarated.

They would not speak to another soul about what they saw that day for twenty years.

This chapter examines potential interactions between the Cro-Magnons and Neandertals in Europe. We know that Neandertals mated with Denisovans and early modern humans, ultimately contributing genes to people today (chapter 6), but much of this gene flow appears to have occurred outside of Europe and prior to the emergence of the European Upper Paleolithic. How frequent were interactions between the Neandertals and Cro-Magnons in Europe after 54,000 years ago? Here the archaeological data have much to say, and we begin by discussing the earliest Upper Paleolithic in Europe.

MANDRIN CAVE AND THE INITIAL UPPER PALEOLITHIC

Some 120 kilometers (75 miles) upstream from where the Rhône empties into the Mediterranean, the cave high upon the hill has a commanding view of the river lying just 5.5 kilometers (3.4 miles) to the west. This is Mandrin Cave, and its location is ideal, positioned along a natural corridor between the Mediterranean and Europe north of the Alps (figure 8.1). Mandrin is famous for its unusual Neronian industry, named for the neighboring site of Néron. At Mandrin the Neronian occupation is sandwiched between Middle Paleolithic levels; it lies atop one Mousterian level and is then covered by five additional Middle Paleolithic layers. Long a curiosity, the Neronian is marked by the production of blades, points, and even microliths less than a centimeter in length called nanopoints. The Neronian level at Mandrin has also yielded an eagle talon featuring cut marks and a worked, almost certainly ornamental, red deer tooth. The technological sophistication of this industry is such that it is characterized as Initial Upper Paleolithic, or IUP—a collection of similar industries spanning Eurasia from Iberia in the west to Mongolia in the east, dating from about 54,000 to 40,000 years ago. In addition to the Neronian, IUP industries in Europe include the Châtelperronian in Spain and France, the Uluzzian in Italy and Greece, the Bohunician and Szeletian in Austria, Czechia, Hungary, Poland, and Slovakia, and the Bachokirian in Bulgaria (see table 2.2). At Mandrin, the mode of manufacture in the Neronian looks nothing like that of the Middle Paleolithic that ante- and postdates it. But the question of who made the Neronian remained a mystery until February 2022.

That month, an amazing discovery was announced regarding Neandertal/Cro-Magnon interactions at Mandrin. A combination of pretreated radiocarbon and luminescence dates demonstrated that the Neronian level there dates to about 54,000 years ago, and a deciduous (baby) tooth from the level is modern,

FIGURE 8.1 European archaeological and paleontological sites (and other locales) mentioned in this chapter.

Source: Map by author.

making Mandrin's Neronian inhabitants the oldest modern humans yet found in Europe.[1] Their presence at Mandrin is short-lived; the 95 percent confidence limits for the dates of the post–Neronian Middle Paleolithic Neandertals show more than 3,000 years of overlap with those of the Neronians. The excavators therefore characterize the Neronian presence at Mandrin as a brief incursion of modern humans into Neandertal territory. What strikes me is that there is *zero* evidence of technological or cultural exchange between the makers of the Neronian and the Neandertals who succeed (or precede) them at Mandrin, and the tools of the two groups are like night and day in form. Also, the Neandertals who replaced the modern humans at Mandrin remain there for 10,000 years, until finally about 44,000 years ago a second Protoaurignacian modern human occupation begins. This is not a simple story of modern humans who, with their brains and superior technology, rapidly outcompeted the Neandertals. In fact, if there is interaction between Neandertals and modern humans at Mandrin (and there may have been none), it is seemingly as fleeting as one could imagine. That said, the fact that the makers of the earliest known Upper Paleolithic in Europe are *H. sapiens* begs the question: Are all IUP industries the work of modern humans, or might some, like the Châtelperronian, be the work of Neandertals? As you will see, the answer to this question is not a simple one.

BACHO KIRO

The Bulgarian cave of Bacho Kiro, on the north slope of the Balkan Mountains, was first excavated by Dorothy Garrod in 1938. Excavations began anew at the site in the 1970s, yielding incomplete remains of fossil hominins (unfortunately later lost), and conventional radiocarbon dates suggesting that the earliest Upper Paleolithic levels were at or beyond the limits of radiocarbon at about 40,000 BP. The site was reopened in 2015 by a joint Bulgarian/German team with the goal of acquiring newer, better, pretreated AMS radiometric dates that would be independently verified by two radiocarbon laboratories. In 2020, this multinational team reported an age for the earliest Upper Paleolithic levels at Bacho Kiro of 46,940 cal BP, at the time making it the oldest Upper Paleolithic in Europe.[2] In addition to finding a human molar, the recent excavations yielded bone fragments that looked as if they might be human. The team also analyzed human-looking, yet undiagnostic, bone fragments that had been recovered during the 1970s excavations. To investigate these, they used new technology known as ZooMS (Zooarchaeology by Mass Spectrometry).[3] This allows one to determine the species of a particular bone or bone fragment by examining the chemical composition of its collagen. ZooMS revealed that these human-looking bone fragments were, in fact, human. Six of the fragments were subsequently radiocarbon dated, and the entire mtDNA genome was reconstructed for five of them and for the tooth. Their mtDNA genomes reveal that all are *H. sapiens*. It is interesting that the molar had a sequence identical to that of one of the fragments, indicating that they either represent the same individual or are maternally related. Another finding is that the Bacho Kiro hominins, among the oldest members of our species in Europe, are more genetically similar to each other in their mtDNA than are 97.5 percent of nonkin Europeans today.[4] They also appear to have living Asian and Native American (but not European) descendants (chapter 6).

Multiple pierced mammal teeth and a single ivory bead were recovered from the earliest Upper Paleolithic levels at Bacho Kiro, as well as several bone awls—all hallmarks of the Upper Paleolithic. In terms of lithics, these tools are called Bachokirian because they had Upper Paleolithic features and blades, along with other more Middle Paleolithic-like implements such as flakes and blades made using Levallois reduction sequences. Hublin et al. point out that this assemblage, like the Neronian, fits comfortably within the IUP.[5]

The results from Bacho Kiro are of great interest because of what fossil and genetic data tell us about the peopling of Europe. Geneticists and paleontologists agree that the earliest members of our species, *Homo sapiens*, evolve first in Africa and are the primary direct ancestors of everyone alive today. When these hominins expanded beyond Africa, however, they encountered the descendants of *H. erectus*, such as the Neandertals and Denisovans, and interbred with them. This is settled science, but we want to know how these modern humans got from Africa to Europe.

The Iberian Peninsula is visible from Africa across the Strait of Gibraltar, a body of water only about 15 kilometers (9 miles) across, but the currents at this 900-meter-deep (2,950 foot) pass between the Mediterranean and Atlantic are so treacherous that even today dozens of would-be immigrants die each year trying to cross it. Thus the southwestern route into Iberia was never a major conduit for population movement in the Pleistocene. Similarly, at many times during the Pleistocene, the Caspian Sea, which was fed by Arctic glacial rivers, extended farther north than today into very cold and inhospitable polar steppe, so population movements into Europe from the northeast were also limited. In contrast, with the lower sea levels of glacial times, there was a land bridge across what is now the Bosporus and Dardanelles, the straits connecting the Black Sea to the Mediterranean. We therefore expect that most modern humans coming into Europe would have entered the continent using this southeastern Balkan route. From an ecological perspective, the date of the IUP at Bacho Kiro is also of interest because, despite its occurrence within a glacial phase, it happens to correspond to a time of slight climatic warming known as Greenland Interstadial 12 (GI12).

OTHER INITIAL UPPER PALEOLITHIC INDUSTRIES

The makers of the IUP from Mandrin and Bacho Kiro are now known to be *H. sapiens*, but it remains uncertain who is responsible for the manufacture of the other IUP industries. There are no diagnostic skeletal remains from either the

Bohunician or the Szeletian in Central Europe.[6] These industries show techno-logical similarities to the local Middle Paleolithic industries that precede them in the region, so many archaeologists suspect their makers were Neandertals.[7] With regard to the Uluzzian and Châtelperronian, however, things are even more complicated.

The Uluzzian is an IUP from Italy and Greece that dates from about 45,000 to 40,000 BP. As with other IUP industries, it is characterized by a significant proportion of artifacts made on flakes, but it shows a higher number of blade tools than is typical of the Mousterian. The most characteristic artifacts of the Uluzzian are crescent-shaped backed bladelets called lunates. Uluzzian lithics are unusual in that many are made with a bipolar reduction technique, in which the core is placed on an anvil, then struck with a hammer, yielding characteristic nonconchoidal fracture scars. The Uluzzian also yields a bone awl, stained with red ochre, at the site of La Fabbrica.[8] Other Uluzzian awls, as well as shell beads, are known from the sites of Grotta del Cavallo (Cave of the Horse), the type site of the industry, and Castelcivita.[9]

Did Neandertals make the Uluzzian? In 2011, Stefano Benazzi and colleagues analyzed the size and shape of two deciduous ("baby") molars from Uluzzian levels at Cavallo and found they fell within the *H. sapiens* range of variation and outside that of Neandertals. Perhaps, then, as at Mandrin and Bacho Kiro, the makers of the Uluzzian are modern humans. Unfortunately, the data are incon-clusive. Zilhão et al. point out that Layer D at Cavallo, from which the teeth are derived, yields multiple Dufour bladelets, small blades with a curved profile that are the hallmark of the Protoaurignacian industry.[10] Thus Layer D could be a Protoaurignacian level or a mixed level containing both Protoaurignacian and Uluzzian artifacts. Of course, if Layer D is Protoaurignacian, and not Uluzzian or a mixture of the two, the bone awls and beads recovered at Cavallo might similarly be non-Uluzzian, as well.[11] For the moment, it's prudent to admit that we don't know who made the Uluzzian.

No IUP industry is as disputed with regard to who is responsible for it than the one from the western edge of Europe, the Châtelperronian. The Châtelp-erronian is found in northeastern Spain and southwest to central France from about 44,000 to 40,000 cal BP. Through at least the 1970s, the Châtelperro-nian was considered a full-fledged Upper Paleolithic industry and was therefore assumed to have been made by Cro-Magnons.

Like other IUP industries in terms of technology, the Châtelperronian has mixed affinities. Its type fossil is the Châtelperron backed knife (chapter 10), and like most Upper Paleolithic tools it is made on a prismatic blade. Indeed, the vast majority (75 percent or more) of Châtelperronian lithics are made on blades. Additional Upper Paleolithic-like aspects of the industry include bone awls, ivory pendants, and pierced mammalian teeth recovered from two sites in central France: Grotte du Renne (Reindeer Cave) at Arcy-sur-Cure and Quinçay. However, a sizable minority of Châtelperronian lithics are made on Levallois flakes and look nearly identical to the Mousterian tools that precede them in the region.[12]

In terms of diagnostic human remains, only at two sites has human skeletal material been recovered from Châtelperronian levels: Grotte du Renne (Arcy) and St.-Césaire. In both cases, the remains are diagnostically Neandertal. At Grotte du Renne, André Leroi-Gourhan (1911–1986), who excavated there from 1948 to 1963, uncovered Neandertal teeth from a basal Châtelperronian level. Many years later, Hublin et al. determined that what had been considered an undiagnostic subadult temporal bone recovered from a Châtelperronian level at Arcy showed Neandertal inner ear anatomy.[13] In 2016, Welker et al. used ZooMS to find that multiple undiagnostic bones from Châtelperronian levels from Grotte du Renne are Neandertal.[14] As the Châtelperronian appears to develop out of the local Mousterian, and given that so far only Neandertals are associated with it, by the 1980s, many scholars, including Leroi-Gourhan, began to view the Châtelperronian as a Neandertal industry, albeit one that may borrow technological know-how from neighboring Cro-Magnons who were making and using Aurignacian tools (more on this soon).

These arguments sound reasonable to me, but almost every sentence I wrote about the Châtelperronian is in dispute. First, multiple archaeologists have questioned the association of the bone awls and beads with the Châtelperronian at the Grotte du Renne, arguing that the Mousterian, Châtelperronian, and Aurignacian layers there have all been mixed. In support of this claim, different teams have generated radiocarbon dates from Grotte du Renne that are out of chronological order with the site's recorded stratigraphy.[15] Ofer Bar-Yosef and Jean-Guillaume Bordes use site records to show that the Châtelperronian residents of the cave smoothed their living floor by excavating and leveling the underlying Mousterian level, even digging postholes into it. They hypothesize that the Neandertal teeth may therefore have come from the underlying Mousterian level.[16]

As for St.-Césaire, Bar-Yosef and Bordes note that the Châtelperronian layer (EJOP sup), which yielded the Neandertal skeleton, lies above a problematic layer (EJOP inf) of unknown affiliation, and Gravina et al. believe it's most likely Mousterian.[17] Bar-Yosef and Bordes also maintain that the discoverers' action of removing the St.-Césaire 1 skeleton en bloc to excavate it in the lab makes the interpretation of its exact position within the site's stratigraphic context impossible. Gravina and colleagues also point out that at most Châtelperronian sites "Mousterian-like" tools are nonetheless made on the by-products of blade production—this is decidedly not the case at St.-Césaire, where the Mousterian-like tools from the Châtelperronian layer are manufactured from more typically Mousterian-like flakes.

Caron et al. counter these critiques, arguing there is no evidence of a high degree of mixing at the Grotte du Renne, given that 100 percent of the site's Levallois flakes come from Mousterian levels, 99 percent of the Châtelperronian points are from Châtelperronian levels, and 100 percent of the bladelets are from a Protoaurignacian level.[18] They dismiss the notion that the beads and awls "migrated" down into the Châtelperronian from the overlying Aurignacian layers because none of the diagnostically Aurignacian lithics appears similarly displaced. Hublin et al. also refute claims questioning the attribution of the Châtelperronian to the Neandertals by generating new, ultrafiltrated radiocarbon dates for the Grotte du Renne—dates that nearly perfectly match the site's stratigraphic sequence.[19] They note that previously dated Grotte du Renne bones were likely contaminated by consolidant applied to stabilize them before removal. Finally, they report an ultrafiltrated calibrated radiocarbon date taken directly on the St.-Césaire 1 Neandertal of 41,950 to 40,660 BP, which neatly corresponds to the timing of the transition between the early to later Châtelperronian layers at Grotte du Renne. In the end, although I side with those who argue the Châtelperronian was manufactured by Neandertals, this conclusion remains far from certain.

THE EMERGENCE OF THE AURIGNACIAN

When I began graduate school in 1988, there was broad consensus that the Aurignacian first developed in the Levant about 42,000 years ago, expanded into the Balkans by around 38,000 years ago, then slowly spread to the west and north, finally reaching western France and northern Spain about 36,000 years ago. Given the geography of Europe, one can see how this hypothesis would appeal to many archaeologists. Then, in November 1989, two back-to-back articles shook the very foundations of this view.[20] I still remember how excited the archaeologist and member of my doctoral committee, Lawrence Straus, was when they came out. Both papers reported seemingly impossibly early uncalibrated AMS radiocarbon dates for the Aurignacian at two Spanish sites: L'Arbreda in Catalonia and El Castillo in Cantabria. At L'Arbreda the 1989 dates had the Protoaurignacian appearing as early as 39.9 ka BP and the Aurignacian proper appearing as early as 37.1 ka BP. At El Castillo, the dates had the Aurignacian proper appearing as early as 39.5 ka BP—calibrated dates would have been older still! Although many researchers initially rejected these unusually early dates, they have stood the test of time, with new dates for the earliest Aurignacian at L'Arbreda at 42.3–40.3 ka cal BP and at El Castillo greater than 42 ka cal BP.[21] Dates this old for the Aurignacian continue to pop up all over Europe.

Currently, the earliest secure date for the Aurignacian proper (not Protoaurignacian) in Europe is about 43,500 cal BP at the site of Willendorf II in Austria.[22] This Danube Valley site is north of the Alps about 51 kilometers (32 miles) SSW of Vienna. Although the date corresponds to a slightly warmer period known as Greenland Interstadial 11, the environment at the time would nonetheless have been a cold open steppe. About 472 kilometers (293 miles) upriver, the earliest Aurignacian levels at Geissenklösterle in southwestern Germany date to about 43,060 to 41,480 cal BP.[23] This is amazingly early because the Feldhofer Neandertal, another 374 kilometers (233 miles) to the NNW, was directly radiocarbon dated to 39,900 ± 620 to 40,394 ± 512 cal BP.[24]

The Mediterranean never experienced the extreme cold that Europe north of the Alps does, but the Aurignacian is not earlier there. The Protoaurignacian at Mandrin Cave dates to about 44,000 years ago. Fumane Cave is in the greater Po Valley just south of the Alps, about 18.5 kilometers (11.5 miles) NNW of Verona. It yields Mousterian, Uluzzian, and Aurignacian layers. Falcucci et al. report new dates for Fumane showing the earliest Uluzzian there dates to about 43,600–43,000 cal BP.[25] This corresponds to the dates for the latest Mousterian at the site, suggesting a rapid transition between these two industries. The Protoaurignacian at Fumane begins some two thousand years later at about 41,200–40,400 cal BP. On the western side of Italy, one of the Balzi Rossi sites, Riparo Mochi, preserves a long sequence from the Mousterian through the Epigravettian. Douka et al. report radiocarbon dates following strict pretreatment protocols for the Protoaurignacian at Riparo Mochi on humanly modified marine shell of about 42.7–41.6 ka cal BP.[26]

What about the western edge of Europe? In 2019, Cortés-Sánchez et al. announced dates for a presumed Aurignacian level at the site of Bajondillo, near Málaga in southern Spain, of 43.0 to 40.8 cal ka BP. This site is controversial, with some arguing the layer in question is mixed and non-Aurignacian in its affinities.[27] Finally, in 2020, the archaeologist Jonathan Haws and his colleagues, including one of my former PhD students, Lukas Friedl, reported dates at the Portuguese site of Lapa do Picareiro for an Aurignacian level sealed from above by a stalagmitic layer at 41.1 to 38.1 ka cal BP.[28] This site sits atop a mountain only 37 kilometers (23 miles) from the Atlantic coast. If the Aurignacian is this far west 41,000 years ago, by this time it could be anywhere in Europe!

In geological terms, the Aurignacian appears to spring up rapidly across Europe about 43,000 to 41,000 years ago (see table 2.2). To me these dates are exciting because if the Châtelperronian lasts until about 40,000 years ago we have somewhere between 1,000 and 3,000 years of time for potential interactions between the makers of the Châtelperronian (Neandertals?) and the makers of the Aurignacian (Cro-Magnons?)—interaction we failed to document at Mandrin Cave 10,000 years earlier.

CHÂTELPERRONIAN AND AURIGNACIAN INTERACTIONS?

I'm sure each of us can think of at least one thing we wistfully wish were true even though it isn't. For me, it's a now-disproven idea about the nature of the early Upper Paleolithic at four sites: Grotte des Fées (Châtelperron),[29] Le Piage and Roc de Combe in France, and El Pendo in northern Spain. It had been widely accepted in the 1990s that at each of these sites there was interstratification of Châtelperronian and Aurignacian levels.[30] If we assume just for the moment that the Châtelperronian was made exclusively by Neandertals and the Aurignacian was made exclusively by Cro-Magnons, this would mean that the same cave or rock shelter was alternately occupied by Neandertals, then Cro-Magnons, and then back to Neandertals again repeatedly over 1,000 to 3,000 years. This "time-share" view of human evolution would be consistent with a long-lived cohabitation of Europe by these two hominin species, one in which they were frequently coming into contact with each other. I can think of few paleoanthropological ideas more evocative than this.[31]

Sadly, in the early years of the twenty-first century, this interpretation of these sites began to fall one by one. First, the archaeologist Jean-Guillaume Bordes reexamined the collections from Roc de Combe and Le Piage for his 2002 dissertation and found that the interstratified sequence at Roc de Combe came from a limited area within the site where there had been mixing of Mousterian, Châtelperronian, Aurignacian, and even Gravettian materials. For the rest of the site, the cultural levels remain in their expected temporal sequence. At Le Piage, through refitting (i.e., piecing cores back together with the flakes/blades and waste products knapped from them) Bordes was able to demonstrate that the Châtelperron knives found interspersed among Aurignacian tools were derived from an area outside of and, more important, above the shelter. These tools had therefore been redeposited into the shelter on top of a pristine Aurignacian level via gravity. Similarly, at El Pendo, the archaeologist Ramón Montes and colleagues showed the entire archaeological sequence was a secondary erosional deposit.[32]

This left only the Grotte des Fées. In 2008, a testy exchange took place in the form of two papers published back-to-back in the journal *PaleoAnthropology*. In the first, Zilhão and colleagues argued there had been extensive disturbance of Grotte des Fées by carnivores—in particular, hyenas—evident from faunal bones showing telltale signs of having been partially digested.[33] Worse still, they claimed that although some of Henri Delporte's excavations in the 1950s and 1960s, the most recent at the site, were into in situ deposits, Zilhão and colleagues held that much of what he dug into was nineteenth-century backfill—the dirt left over after an excavation that is then shoveled back into the site to refill the hole. In a response paper, Paul Mellars and Brad Gravina took exception to this claim, calling it an ad hominem attack from researchers blinded by theoretical bias.[34] They argued that an archaeologist as capable and experienced as Delporte surely would have recognized when he was digging into backfill.

Later that same year the Canadian archaeologist Julien Riel-Salvatore and colleagues tried to stake out the middle ground between these camps.[35] In a contribution to *World Archaeology*, they pointed out that the discovery of Aurignacian-looking artifacts in "good" Châtelperronian levels is not a rare occurrence. Ultimately, however, they concluded that none of the reported Châtelperronian-Aurignacian sites shows unequivocal evidence of interstratification of these industries.

Why does the potential for Châtelperronian-Aurignacian interstratification at a site engender such impassioned arguments? It boils down to the debate on modern human behavior (chapter 7). Paul Mellars has long maintained that the Châtelperronian is the result of Neandertals imitating the material culture, including personal ornamentation, of the neighboring Aurignacians, without fully comprehending its significance.[36] He has repeatedly said it is far too coincidental that Neandertals do not produce personal ornaments until we find Cro-Magnons in Europe. For this imitation scenario to make sense, however, we need a sustained period of Neandertal and Cro-Magnon coexistence and interaction within the same region—not merely the brief incursion we saw with the Neronian level at Mandrin.

In contrast to Mellars, Zilhão has long argued that Neandertals invented the Châtelperronian on their own, without Cro-Magnon influence.[37] He and his colleagues have claimed that for every site at which the Châtelperronian and Aurignacian co-occur, one cannot reject the hypothesis that the Châtelperronian is older than the Aurignacian. They also argue that in Europe one cannot reject the hypothesis that the Aurignacian is younger than 36,500 radiocarbon years BP (about 38,500 cal BP).[38] Given the multitude of recent ultrafiltrated Aurignacian dates that are greater than 40,000 cal BP, this latter hypothesis is now rejected. That said, although we may not see temporal overlap between the

Châtelperronian and Aurignacian at any one site given the ranges for radiometric dates available for the two industries, I believe we cannot reject the hypothesis of at least one or two millennia of coexistence of these two industries. Again, if the Châtelperronian is exclusively the work of Neandertals and the Aurignacian the work of the Cro-Magnons,[39] it suggests that these two hominins overlapped in the region for at least a thousand years.

In the end, my take on the "imitation vs. independent invention" debate is that it is much ado about nothing. I base this on an unusual parallel—the issue of whether wild chimpanzees engage in cultural behavior. In a 1992 book that has shaped my thinking for decades, the primatologist William McGrew examined the behavior of wild chimpanzees across Africa using a set of six criteria that the American anthropologist Alfred Kroeber (1876–1960) came up with for recognizing cultural acts.[40] To Kroeber's list McGrew adds two more criteria he deemed necessary for determining whether a behavioral act of a *nonhuman* species qualifies as culture. The resulting list of eight criteria for recognizing behaviors as cultural, along with their definitions, is presented in table 8.1. For the Early Paleolithic, I would argue that even the 3.3-million-year-old Lomekwian tools meet all but one of these criteria (nonsubsistence: the behavior is not directly related to the procurement of food). Note that the eighth criterion, naturalness (the action is performed in the absence of human interference), would not apply to this case. Reflecting on the archaeological evidence presented in chapter 7, I would argue that by the Middle Paleolithic hominins are engaging in behaviors

TABLE 8.1 Criteria for determining whether a behavioral act is cultural

1   Innovation: new pattern is invented or modified.

2   Dissemination: new pattern acquired by another from innovator.

3   Standardization: form of pattern is consistent and stylized.

4   Durability: pattern performed without presence of the demonstrator.

5   Diffusion: pattern spreads from one group to another.

6   Tradition: pattern persists from innovator's generation to the next one.

7   Nonsubsistence: pattern transcends subsistence.

8   Naturalness: pattern shown in absence of direct human influence.

Sources: Numbers 1–6 are from Alfred L. Kroeber, "Sub-Human Cultural Beginnings," *Quarterly Review of Biology* 3 (1928): 325–42. Numbers 7 and 8 were added by William C. McGrew, *Chimpanzee Material Culture: Implications for Human Evolution* (Cambridge: Cambridge University Press, 1992).

that meet the first seven criteria. It is important to note that McGrew found that no single regional group of wild chimpanzees exhibits behaviors meeting all eight criteria, but examples of all eight behavioral-cultural criteria can be found across the species as a whole.

A detailed reexamination of these criteria is beyond the scope of this book, but in thinking of the "imitation vs. invention" Châtelperronian debate, I come back to McGrew's criterion on naturalness (a behavioral pattern that occurs in the absence of direct human influence). When I first read McGrew's book, I reflected on the Taï Forest chimpanzees in Côte d'Ivoire who use hammer-stones and anvils to break open nuts. My first thought was that it is impossible for us to know whether the first chimpanzee who performed this nut-cracking behavior did so after surreptitiously watching humans doing the same thing, but my second thought was to wonder whether this is even a relevant consideration. After all, the first human to invent any particular cultural behavior is the only one who invents it. In the absence of independent invention, everyone else either imitates the innovator, learns the behavior from the innovator, or learns it from someone who was taught by someone else.

In the same light, short of finding 100,000-year-old pierced teeth and shell beads in a European Mousterian site, we will never know if the Neandertals copied these elements of material culture from the Cro-Magnons or invented them themselves.[41] But as discussed in chapter 7, Neandertals appear to have had the *capacity* for modern human behavior, even if that capacity is only occasionally expressed in the Mousterian. Remember that the actual expression of "modern human behavior" appears to be context-dependent for *H. sapiens* as well—there are brief periods of technological and artistic florescence in the African MSA (e.g., the Howiesons Poort and Still Bay industries) that appear for a few thousand years and subsequently disappear. Their disappearance is not due to some cognitive defect but rather is almost certainly in response to an environmental factor, perhaps one as simple as decreased population density—and carnivore ecology informs us that Neandertals lived at low population densities indeed!

MOUSTERIAN AND AURIGNACIAN INTERACTIONS IN IBERIA?

Iberia south of the Cantabrian Mountains has long been thought a locale for late interaction between Neandertals and Cro-Magnons. The British archaeologist Clive Finlayson argues that prior to 29,000 years ago much of this region would have been Mediterranean wooded savanna—a relatively open habitat expected to show elevated richness in mammalian species.[42] As such, southern and, to a lesser extent, central Iberia was thought to have served as a refugium for the last Neandertals. By way of contrast, the Cro-Magnons, sweeping across the continent from the east, outcompeted the Neandertals in other colder parts of Europe, including Spain north of the Cantabrian Mountains, which today is known as "Green Spain" because of its cooler, wetter weather. Zilhão went so far as to put forward what he called the Ebro Frontier model, in which Iberia south of the Cantabrian mountains and west of the Ebro River was a Neandertal refugium not colonized by the Cro-Magnons until the onset of the late Aurignacian.[43] Finlayson agrees, arguing the replacement of Neandertals by Cro-Magnons in southern Iberia is the result of climatic cooling around 29,000 BP, transforming much of the region to a dense montane forest, an environment he claims Neandertals never exploited.[44]

For many years, the archaeological record and radiometric dates from southwestern Iberia (in Spain, Portugal, and Gibraltar) seemed to support the Ebro Frontier view. Specifically, there seemed to be an absence of Aurignacian o or I in southern Iberia, whereas the Mousterian there seemed to survive very late. Two sites in Andalusia—Zafarraya and Carihuela—were among these late Mousterian sites. The Mousterian at Zafarraya, which yielded a Neandertal mandible, was dated to 27.0 ka BP via U-series and 29.8 ± 0.6 ka BP (uncalibrated) via radiocarbon.[45] Uncalibrated radiocarbon dates of the terminal Mousterian at Carihuela Cave were about 28,440 and 21,430 BP.[46] Similarly, uncalibrated dates on charcoal from the uppermost Mousterian levels at Gorham's Cave in Gibraltar were reported at about 24,010 ± 320 BP to 32,560 ± 780 BP. The younger of these Gibraltar dates is almost certainly erroneous because it comes from the back of the cave where a Mousterian and Solutrean layer come into direct contact with each other. But the older date, taking error ranges into account, would suggest Neandertals survived there to about 36,000 cal BP.[47] At Jarama, in the La Mancha region of central Spain, the Mousterian was dated to as recently as 26.9 to 29.4 cal ka BP.[48] Finally, at Gruta da Oliveira, Portugal, Angelucci and Zilhão reported uncalibrated radiocarbon dates on burnt bone for Mousterian layer 8 of about 35 to 38 ka BP.[49] Lying atop layer 8 is yet another Mousterian layer that was not dated, but presumably it would be younger still.

Here, too, recent radiocarbon dates using more rigorous pretreatment protocols are providing a different picture of the timing of the end of the Mousterian (and therefore presumably the last Neandertals) over much of Iberia. First, for Jarama, ultrafiltrated radiocarbon dates yielded an age for the late Mousterian of about 50 ka cal BP or older—dates consistent with luminescence ages suggesting the late Mousterian at the site is around 50,000 to 60,000 years old.[50]

At Zafarraya, when a bone that had previously given an uncalibrated date of 33,300 BP was redated using ultrafiltration, an uncalibrated date of >46,700 BP was obtained, but at Carihuela not a single sample was found with sufficient organic material for radiocarbon dating![51] As Haws et al. note, at this point the only late (i.e., younger than 37,000 cal BP) Mousterian dates left in the Iberian Peninsula are Gruta da Oliveira, Cueva Antón, and Gorham's Cave—dates that now stick out like a proverbial "sore thumb."[52]

Don't get me wrong; it makes sense that Neandertals could survive late in the southern (and especially coastal) reaches of Iberia, but the old dates showing them surviving as recently as 32,000 years ago are now anomalous. Across Europe (including almost all sites in the Mediterranean Basin) Neandertals seemed to disappear about 40,000 years ago. In this light, new radiocarbon dates using pre-treatment protocols are desperately needed for these very late Mousterian levels, especially those in Gibraltar, to verify whether Neandertals truly did survive to around 32,000 years ago or later there. The previous dates imply 9,000 years of cohabitation between Neandertals and Cro-Magnons in Iberia.[53]

Nonetheless, between 54,000 and 40,000 years ago, the Cro-Magnons and Neandertals almost certainly repeatedly met each other in Europe. Given the large amount of time and territory over which these interactions occurred, they may have ranged from aggression and outright violence to mutual avoidance, and from grudging tolerance to friendly and hospitable. These interactions occasionally involved mating as well. Although I wish it otherwise, and despite the long temporal overlap, I suspect contacts between Neandertals and Cro-Magnons in Europe after 50,000 years ago were rare for two reasons. First, I am swayed by Churchill's work suggesting that even before modern humans expanded into Europe Neandertals were already living at very low population densities. Second, as of this writing, none of the late (i.e., < 50,000 BP) European Neandertals from whom we have been able to extract DNA shows any signs of interbreeding with modern humans. In contrast, some of the earliest Cro-Magnons do show signs of then-recent Neandertal ancestry, although there also appears to be selection against Neandertal genes in modern humans (chapter 6). In the next chapter. I return to a more bioanthropological approach to the Cro-Magnons themselves and explore their paleobiology.

9
Bioanthropology of the Cro-Magnons

T his chapter begins with a discussion of so-called biological races, a problematic paradigm long a mainstay of biological anthropology. The first bioanthropological studies of the Cro-Magnons in the late nineteenth and early twentieth centuries spilt a lot of ink assessing the "racial affinities" of these prehistoric humans. In this chapter, you will become reacquainted with many old fossil friends, meet new ones, and learn what their skeletons, the ancient DNA drawn from them, and their burials tell us about them. Finally, I examine how prehistoric behavior affects the skeleton using an example from the upper limb.

THE PROBLEM OF BIOLOGICAL RACE

Biological anthropologists are not particularly proud of our history as a discipline, especially in its earliest days. But even through the 1970s many biological anthropologists were obsessed with the concept of biological races. As a result, countless bioanthropological studies conducted through the mid-twentieth century were blatantly and unapologetically racist. Worse still, in the 1920s and 1930s, bioanthropologists on both sides of the Atlantic were heavily involved in the eugenics movement, which originated in the United States. Eugenicists argued that certain groups or types of people—people of color, those deemed to have mental illness or to be mentally inferior, and many other socially marginalized people—should not be allowed to reproduce to prevent the "degradation" of the white gene pool. This abhorrent view led to mass involuntary sterilization in the United States and played a role in the "Final Solution" in Nazi Germany; in addition to his MD, Josef Mengele had a PhD in biological anthropology.

Bioanthropology today has thankfully exorcised these metaphorical demons. However, important issues surrounding race are still very much debated in the public square, and race is an area in which bioanthropologists have much to contribute. I'll tackle this thorny topic using several thought experiments, the first of which is this:

> Imagine you are walking down a crowded city sidewalk. Your senses are bombarded; you hear sirens and car horns, the revving of diesel engines. You smell exhaust and the rich aroma of food from street vendors. And you see people, lots of people. You could be in Mexico City, Sydney, Johannesburg, London, Paris, or New York—anywhere you like. As you stroll, you encounter a sea of people, and you notice all sorts of faces and hair types and skin colors. You may even catch yourself wondering about the ethnicity or geographic origins of some of the people you pass.

But I know that you, gentle reader, would never be so rude as to inquire about their ancestry! Our thought experiment in this case is a simple one—how accurate are your assessments of the ancestry of these strangers? Although you are correctly guessing the broad ancestry of at least some of the people you see, you are probably less accurate than you realize because your own life experience greatly influences your perception of human geographical diversity.

To illustrate this point, I'll tell you another story from my time collecting dissertation data. In February 1994, I was in the bucolic setting of Cambridge University. There are no fossils for me to study here. Instead, I'm collecting comparative data—measuring recent (the last 5,000 years or so) human skeletons to which I will compare my Pleistocene fossils. We paleoanthropologists are frequently taken to task by scholars in other disciplines for our small sample sizes. The paleoanthropologist Gary Schwartz once told me that he published a paper on 1/32 of an individual—you guessed it: he wrote a paper on a single tooth! The fossil record is what it is, and fossil samples will always remain woefully tiny because of the rarity that any living organism actually becomes a fossil. One thing we can control, however, is the size of our comparative data sets, and this is what brought me to Cambridge. I am working with a large group of skeletons known as the Duckworth Collection, and during my stay I am not alone. A senior Australian researcher, now long retired, is also working here. He is examining crania; I am studying postcrania. I seek skeletons that have preserved vertebral columns. He helps me by announcing which specimens he's found that have backbones,

and I let him know about any skeleton I happen to come across that has a particularly well-preserved cranium. It's nice to have the company and the camaraderie.

One afternoon as we are up to our elbows in skeletons (one does get used to
this, by the way), we are visited by a colleague who, for discretionary purposes,
shall remain nameless. At his side is a young woman who appears to be in her early
twenties. She doesn't say anything to either of us. It is the professor, many years
her senior, who does all the talking (go figure). After they leave, the Australian
turns to me and says, "She's got an exotic look. I wonder if she's Maori?" What
immediately struck me is that I would never assume *anyone* I met was Maori (an
Indigenous group of New Zealand)—it's just not part of my lived experience as
an American. I told him I had assumed she was Latina. In contrast to Maoris,
Latin Americans are an everyday part of my lived experience in the United States.

I found out some weeks later that I was correct (she was Latina), but my being
right was purely by chance. It was not because I had a better eye than my Australian colleague for human variation. When we evaluate other people's ancestry,
our own experience plays a huge role in how we perceive it.

It is also important to recognize that races can be either sociocultural or biological. Sociocultural races are real because they have real consequences for real
people in real life. I shudder when I hear some academics say that races do not
exist. For me, this is the liberal equivalent of Stephen Colbert's conservative character when he avers that he "does not see color." It is also easy for a white person
in the United States to proclaim that races do not exist. It is much more difficult
for people of color who have faced real racial discrimination all their lives to
make such a claim.

Sociocultural races are real, at least in their consequences, but biological races
are not. This may be a hard pill for some of you to swallow, for a host of reasons:
(1) many of us, especially older Americans, were taught at an impressionable
young age that biological races are real; (2) the reality of sociocultural races can
easily be transferred to biological races if one only thinks about these issues in
a superficial manner; and (3) humans obviously vary biologically with geography in well-known ways, many of which are associated with a few highly visible
characteristics that we as humans tend to notice, such as skin color and facial
features. Humans' ability to recognize faces is uncanny, and it has been fashioned
by natural selection. Like most primates, humans are highly visual animals who
see in color, and this increase in visual acuity is associated with a decrease in acuity in other senses. An obvious example is that primates lack the amazingly acute
chemoreceptive nose of dogs, which are said to be 10,000 to 100,000 times more
sensitive than our own. As I tell my students, chimpanzees don't smell any better
than we do! If dogs had a race (or species) concept, I believe they would identify

them via olfactory rather than visual cues. It has always amazed me that a Great Dane will recognize a Chihuahua as a dog. If dogs based their judgments on who is a dog on how the dog looks, would they draw the same conclusions?

Time for our next thought experiment:

> Imagine three groups of ten people, all of whom are wearing identical clothes (you can give them any outfit you like). The first group of ten are from Iceland, and their ancestors came to that island from Norway in the ninth century. The second group of ten people are from the Benga language group in Gabon, who live on the Atlantic coast of Africa just north of the equator. And the third group of ten are ethnic Koreans living in Japan. How well do you think you would do if I asked you to sort another reader's thirty individuals to the continent of their ancestors, giving you the choice of Africa, Asia, or Europe?

Unlike the previous thought experiment, in which I expected you to make at least some mistakes, here I believe you would get 100 percent of the assignments right—which is why we have what seems to be a paradox.

If we can so accurately assign people to geographical groups this way, based solely on a few visible aspects of their biology—not the clothes they are wearing or other cultural cues—then biological races must exist, right? Perhaps there are three human biological races, and I've just given you an appropriate example from each. Not really. You see, I didn't pick those three locations at random; I was following the so-called three-race scheme (Caucasian/white, Ethiopian/Black, and Mongolian/Yellow) put forward by Georges Cuvier in his 1817 book *Le Règne Animal* (*The Animal Kingdom*).[1] In framing his three-race scheme, Cuvier may have drawn upon the biblical story of Noah's three sons, who in the tenth chapter of Genesis are said to have moved in different directions after the flood to repopulate the Earth. Cuvier was, after all, a devout Protestant who rejected evolutionary ideas his entire life. Since the earliest attempts to define biological races of humans in the seventeenth and eighteenth centuries, the problem with positing *any* number of races is that there has *never* been agreement on how many biological races of humans exist, or how these races should be circumscribed.

Johann Friedrich Blumenbach (1752–1840), a German physician and anatomist who amassed a large collection of skulls, is widely considered to be the "father" of biological anthropology. In the 1795 edition of his book *De Generis Humani Varietate Nativa* (*On the Natural Variety of Mankind*), he proposes the existence of *five* human races (Ethiopian, Caucasian, Mongolian, Malayan, and American).[2] These are the people of sub-Saharan Africa, Europe and the Mediterranean, northern Asia, southern Asia, and the Indigenous Americans, respectively. According to Blumenbach, these groups blend into each other rather than showing discrete boundaries. Blumenbach is also the first person to refer to the people of Europe, Western Asia, and North Africa as "Caucasians." He did so for two reasons: (1) He thought Europeans' origins were in the Caucasus Mountains—as was the case with Cuvier, this might be because Noah's Ark is said to have come to rest on Mount Ararat, a dormant volcano just south of the Lesser Caucasus Range; and (2) what he deemed the most beautiful skull in his collection was from Georgia, a nation on the southern slopes of the Greater Caucasus. Knowing Blumenbach's story, it bothers me when people use the pseudoscientific term "Caucasian."

Fast-forwarding to the twentieth century, the Harvard bioanthropologist Earnest Hooton (1887–1954) defined a biological race as "a great division of mankind, the members of which, though individually varying, are characterized by a certain combination of morphological and metrical features, principally non-adaptive, which have been derived from their common descent."[3] Jeffrey Long refers to this as the "essentialist" definition of race.[4] I might call it the "Potter Stewart" definition of race, after the Supreme Court Justice who in 1964 refused to define what pornography was, arguing instead that "I know it when I see it." Hooton's definition is, I think, purposefully vague because he knew how messy human biological variation is. He also argued that there were no "primary" (read "pure") races left because groups of humans had been mixing with each other for millennia. Nonetheless, despite their mixed status, Hooton also maintained that these so-called secondary races could still be identified biologically; he recognized no fewer than eighteen of these.[5] Note, too, that at that time Henri Vallois, across the Atlantic, held a concept of biological race nearly identical to Hooton's.

Another twentieth-century Harvard anthropologist, Carleton Coon (1904–1981), thought his *Origin of Races* (1962) would be the most important book in biology since Darwin's *Origin of Species*. To twenty-first century eyes, however, it is both racist and antiquated in its biology.[6] Coon equates human races with biological subspecies (I'll explain what these are in a bit), but his five subspecies are *not* the same as Blumenbach's five races. Instead, Coon argues that the five (rather unfortunately named) subspecies of humans are Caucasoid, Negroid, Capoid, Mongoloid, and Australoid. In Coon's view, Caucasoids were the people of Europe, North Africa, the Middle East, as well as the majority of inhabitants of the Indian subcontinent. The unfortunately named Negroids were sub-Saharan Africans, except for southernmost Africa, whose native people he called Capoid, after the Cape of Good Hope. Mongoloids were the people who lived across most of the huge Asian landmass; for Coon, this unfortunate appellation also encompasses Native Americans and most Pacific Islanders. Australoids were the people of Australia, Tasmania, and New Guinea, with additional inexplicably isolated pockets of Australoid people on the Indian subcontinent, the Andaman Islands, the Malay Peninsula, and the Philippines.

In contrast to Coon's five subspecies, the bioanthropologist Stanley Garn (1922–2007) recognized nine major, or geographical, races: (1) American Indian, for the Indigenous people of the Americas (except the Inuit and Aleuts); (2) Asian, the people found across the bulk of the largest continent, plus Inuit and Aleuts in North America; (3) African, the Indigenous people south of the Sahara; (4) European, including the Middle East and much of North Africa; (5) Indian, for the inhabitants of the Indian subcontinent; and (6) Australian, for the Indigenous inhabitants of that continent (including Tasmania). It is surprising to note that in the Pacific Garn finds three major races: (7) Melanesian, including the people of New Guinea; (8) Micronesian; and (9) Polynesian.[7]

There are no objective criteria by which to reject or accept either Coon's or Garn's models (or any other racial scheme, for that matter). Of what use is a race concept if one cannot objectively determine how many races there are or where their boundaries lie? As Blumenbach realized in the eighteenth century, all boundaries between presumed biological races are arbitrary. I illustrate this point with another thought experiment:

Imagine groups of ten individuals each, all dressed in identical clothing. First, imagine ten Sicilians. Now imagine ten Greeks. How well could you tell these two groups apart? Now imagine ten Turks. Can you reliably distinguish the Turks from the Greeks? Picture ten Syrians. How readily could you distinguish them from the Turks? If you imagine ten northern Iraqis, are they distinguishable from the Syrians? Are ten southern Iraqis easily distinguished from ten northern Iraqis? Are those southern Iraqis themselves distinguished easily from southwestern Iranians? How well can you discern ten people from southwestern Iran from ten eastern Iranians? Can you distinguish ten eastern Iranians from ten people from just across the border in western Pakistan? How easily could you distinguish the ten people from western Pakistan from ten people from eastern Pakistan, and could you reliably distinguish those ten eastern Pakistanis from ten people just across the border in Rajasthan, India? Finally, how reliably would you be able to distinguish those ten western Indians from ten people from southwestern India?

In each case so far, I would argue that you would have little, if any, success in discriminating these groups from neighboring regions.

But could you distinguish our original ten Sicilians from the ten people we just met in southwestern India? I believe you could do this with a high degree of accuracy. But where, exactly, along the way did we cross a racial boundary? The truth is that no such boundary is discernible. Perhaps you, like Coon, think that all the people in our thought experiment are Caucasians, in which case no racial boundaries have been crossed. Now conduct the same thought experiment, but this time from Syria turn to the south through Jordan, Egypt, Sudan, South Sudan and Kenya. On average, skin color will get darker over this journey, but finding an actual racial border will remain a challenge. What we have in each of these examples are *clines*—geographical gradients in measurable characters—but we have no racial boundaries.

Beginning in the 1960s, there was a shift away from an emphasis on these types of anatomical definitions of race to a closer examination of human genetic differences across the globe. In a classic 1972 contribution, the geneticist Richard Lewontin (1929–2021) looked at a wide array of human genetic data and demonstrated that the bulk (over 85 percent) of genetic differences among humans were manifest at the local subpopulation level, and only about 6 percent were attributable to differences between continental (racial) groups.[8] Although there was biological variation within a race, Hooton, Coon, Garn, and other defenders of biological races had argued that members of a particular race were nonetheless more closely related to each other and shared more features with each other than they did with members of another race. Lewontin's pioneering work, and the subsequent explosion in genetic analyses over the past few decades, has shown that this is simply not the case. For example, Jeffrey Long and colleagues used a clustering algorithm to determine which geographical human groups are more closely genetically related to each other. A fully hierarchical model allows populations to be nested independently from any preconceived notion of race and is a significantly better fit for actual genetic data than one constraining populations to be members of races who are more genetically similar to each other than they are to members of other races (figure 9.1).[9]

I had earlier promised to discuss subspecies. In the 1758 edition of *Systema Naturae*, Linnaeus split humans into subspecies, but in zoology, the first systematic use of subspecies came about at the end of the nineteenth century, when field biologists discovered that at times species freely hybridized with each other in areas where their ranges overlapped. To many scientists, this meant they were no longer separate species but were instead subspecies of a single species. A paleoanthropological example of this phenomenon would be modern humans and Neandertals. We know from DNA that Neandertals successfully interbred with modern humans, so some consider Neandertals to be members of our species.

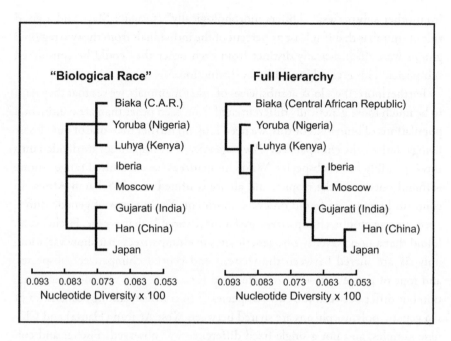

FIGURE 9.1 Cluster diagrams of recent human DNA sequence data. The cluster on the left constrains populations to be members of biological races; the cluster on the right has no such constraint. The model on the right is a statistically significant better fit to the data.

Source: Redrawn from Jeffrey C. Long, Jie Li, and Meghan E. Healy, "Human DNA Sequences: More Variation and Less Race," *American Journal of Physical Anthropology* 139 (2009): 23–34.

When species are reduced to subspecific status, taxonomists have rules about which species should be sunk into the other. In zoology these rules are codified in the International Code of Zoological Nomenclature. One such rule is the priority rule, which states that the earliest species name to appear in the scientific literature has priority over later names. The taxonomic name *Homo sapiens*, having been coined by Linnaeus in 1758, therefore has taxonomic priority over *Homo neanderthalensis*, proposed by King in 1864. As such, Neandertals must become a subspecies of *H. sapiens*, not the other way around. Subspecies are designated by expanding the Linnaean binomial into a trinomial. Thus, in the subspecific scheme, if modern humans and Neandertals are conspecifics, the Neandertals are *Homo sapiens neanderthalensis*; and to distinguish us from Neandertals, modern humans must become *Homo sapiens sapiens*.

There is no agreement in biology regarding when a regional variant of a species warrants subspecific status. Ernst Mayr defined a subspecies as a regional variant of a species that occupies part of the species' range and differs morphologically from other subspecies.[10] "Differs morphologically" is vague; Mayr said a good rule of thumb is that if at least 75 percent of the individuals from the two regional groups were diagnostically distinct from each other they could be considered subspecies. This is clearly an arbitrary distinction.

Furthermore, if we look at subspecies of other mammals, we see that they tend to be much more genetically differentiated from each other than are continental populations of humans. For this, we need look no further than one of our closest living relatives, the chimpanzees (*Pan troglodytes*). Chimpanzees are divided into three, possibly four, subspecies. Wild chimpanzees have a limited range—none is found outside of the tropics, and all are confined to sub-Saharan Africa. In contrast, humans are found on every continent. Yet in terms of genetic differences among groups, chimpanzees are far more variable than we are. Fischer et al. found that among seventy-one genetic sites in chimpanzees that show variation, only six are shared between the western and central chimpanzee subspecies, and four of the seventy-one sites are fixed (i.e., show zero within-group variation but differ between the two subspecies).[11] In contrast, among humans, 57 of 139 genetic polymorphisms are shared between West African (Hausa) and Chinese samples, and not a single fixed difference was observed! Fischer and colleagues also point out that one subspecies, the Central African chimpanzee (*Pan troglodytes troglodytes*) is about 2.4 times as genetically diverse as humans are worldwide! This is even more surprising when one considers that this subspecies has a range of about 700,000 kilometers^2 (270,000 square miles, roughly the size of Texas). In the end, the small genetic differences observed among continental groups of humans do not warrant subspecific status. Thinking about the fossil record, the low genetic variability of humanity today suggests that the visible "racial" features on which we tend to focus may have only recently emerged.

THE "CRO-MAGNON RACE"?

After learning the history of the biological race concept, it will not surprise you to learn that scientists doing the earliest studies of Cro-Magnon fossils were obsessed with these skeletons' racial affinities. With regard to the Cro-Magnon Race, Broca was perplexed by the juxtaposition of the "highly evolved" brains of Cro-Magnon 1 and 2 in the context of more massive (read "primitive") faces and bodies. René Verneau actually argued for two races of *Homo sapiens* from the Balzi Rossi, a more evolved Cro-Magnon Race and a more primitive Grimaldi Race.[12] We have seen what a fool's errand racial classification is with living, breathing human beings. Imagine doing the same with the skeletons of people who lived 30,000 years ago! That said, what would a fleshed-out Cro-Magnon look like? We already know that early Cro-Magnons tended to have longer limbs relative to their trunks than do people in Europe today, which I argue is indicative of their then-recent ancestry from Africa. This begs the question—might they also have had dark-colored skin? If we saw them walking down our hypothetical city sidewalk, what might we guess about their ancestry? As with many paleoanthropological issues these days, a surprising answer to this question has come from the study of DNA.

There are now multiple skeletons from the European Mesolithic hunter-gatherers dating to about 10,000 to 6,000 years ago from which scientists have been able to extract high-coverage ancient DNA. These include Gough's Cave (England), Loschbour (Luxembourg), La Braña (Spain), and Motala (Sweden) (figure 9.2). What is fascinating is that analyses of their DNA show all four of these individuals had dark brown to blackish-colored skin, brown to black hair, and blue to green colored eyes! A facial reproduction of one of these specimens is depicted in figure 9.3. In addition, a penecontemporary specimen from Tiszaszőlős-Domaháza, Hungary, at about 7,600 cal. BP, is culturally associated with Neolithic farming, but his genes link him to earlier hunter-gatherers from Europe (he is a member of the Villabruna cluster that also includes some Cro-Magnon skeletons; chapter 6). As with the four Mesolithic skeletons, in life he, too, was characterized by dark skin and hair, along with blue eyes. Given that the light skin allele for the SLC24A5 gene is almost completely fixed among people of European ancestry today, I doubt many of us would immediately identify these dark-skinned individuals as Europeans were we to encounter them in a modern setting. Although the lighter-skin allele for SLC24A5 is not present in most of the above specimens, it is interesting that the Motala individual had one copy of the light-colored allele he inherited from one of his parents, showing the European Mesolithic population was polymorphic for this gene.

DNA analysis of living people also provides insight into the evolution of skin color. In 2007, Heather Norton and colleagues showed that variation in three genes—SLC24A5, MATP, and TYR—are implicated in the evolution of lighter skin in Europe, but the lighter skin of people in East Asia is due to selection favoring polymorphisms at different genes.[13] In addition, recent work on Native American genetics has revealed evidence for selection favoring lighter skin in their Beringian ancestors. In 2014, Sandra Wilde and colleagues used ancient and recent human DNA to demonstrate that there was a selective sweep strongly favoring light skin via three genetic pathways (SLC45A2, HERC2, and TYR) in Europe about 5,000 years ago.[14] This broadly corresponds to the time of

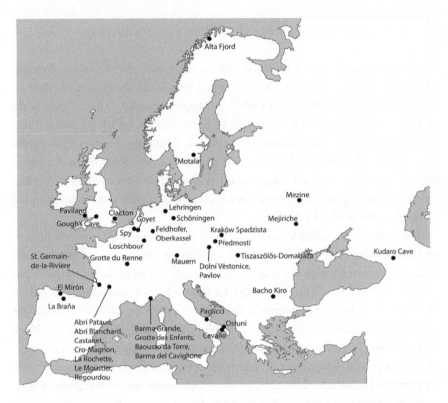

FIGURE 9.2 European locales and archaeological and paleontological sites mentioned in chapters 9 and 10.

Source: Map by author.

FIGURE 9.3 Facial reproduction of the Gough's Cave Mesolithic skeleton by the Kennis brothers, based in part on genetic data.

Source: Courtesy of Alfons and Adrie Kennis.

widespread adoption of agriculture on that continent, and it is thought that a change in diet may have brought about what, in evolutionary terms at least, is the late fixation of light skin in Europe. To explain how this occurred, some background on the evolution of skin color is warranted.

You may be surprised to learn that humans, chimpanzees, and bonobos have approximately equal numbers of hair follicles per unit area of skin. The difference is that except in a few key places—the head, armpits, and groin—human hairs tend to be tiny, wispy bits of nothing. Once our early hominin ancestors had shed the hairy bodies of our apelike ancestors, perhaps with the emergence of *Homo* or *H. erectus* (although it could have been earlier), there would have been selection in Africa and lower latitudes outside of Africa for darker skin. Darker skin is the result of the action of cells deep within the epidermis called melanocytes. Melanocytes produce melanin, an ultraviolet light-blocking brown pigment; think of it as natural sunscreen. More heavily melanized skin protects humans from sunburn, skin cancer, and photolysis, the UV light-induced photochemical breakdown of important substances in the blood such as folic acid.

Why might lighter skin be favored later in a high-latitude environment? This is due to the one positive aspect of exposing the body to UV light—the manufacture of vitamin D. You've heard people refer to skin as the body's largest organ. It is also an endocrine gland, using UV light to synthesize a precursor of vitamin D within your skin's blood vessels. If you have dark skin and live in a cold environment, much of your time outdoors is spent with your skin covered by clothing. Add to this that at high latitudes there are many months with very few sunlit hours and you could end up with vitamin D deficiency, known as rickets in children and osteomalacia in adults. Rickets is selectively significant because it leads to malformations of the limbs and pelvis, and those of the pelvis in particular could have deadly consequences for women during childbirth. This health risk for dark-skinned people living in the north was discovered via a "natural experiment" in the 1920s, when African American children in northern cities such as Chicago and Detroit suffered high rates of rickets. In the end, this public health crisis was solved by adding vitamin D to milk, and fortified milk remains an American staple to this day.

So how did the Cro-Magnons fare so well with dark skin at such high latitudes? Here's the interesting thing about our dark-skinned Cro-Magnons: as long as your diet includes sufficient amounts of vitamin D, you'll have no trouble living in a high latitude environment with deeply pigmented skin. The Inuit people of Siberia, Alaska, Arctic Canada, and Greenland get all the vitamin D they need from the aquatic resources they eat; their skin is not that lightly pigmented. What we therefore imagine happening in Europe about 5,000 years ago is that the descendants of the Cro-Magnons were adopting agriculture and changing to a diet with insufficient vitamin D. In this situation, we expect selection to favor lightly pigmented skin because having even a little bit of lightly pigmented skin—the face or hands, for example—exposed to brief bouts of sun should stimulate the manufacture of sufficient amounts of vitamin D, thereby preventing rickets. In any case, the fact that the lighter skin of Europeans only dates to around 5,000 years ago is an amazing and surprising discovery.

SKELETON KEYS

Human osteology, the study of the human skeleton, is key to the paleoanthropological enterprise. You have already seen how skeletal morphology is used to circumscribe *H. sapiens* and to address modern human origins. Now you will learn about the reconstruction of life from ancient skeletons. Paul Broca, the first bioanthropologist to study Cro-Magnon 1 (CM1), noted that the specimen's mandibular ramus, where the chewing muscles attach, was wider than any human ramus he had ever measured; and at first glance, CM1's face appeared short to him. However, via comparison of the specimen to 123 human crania, Broca found that its facial height falls near the recent human mean, and only one recent human skull has a facial breadth as wide as that of CM1.[15] So CM1's face is not short; it just *looks* short because it's *wider* than those of most people today.

One thing Broca likely gets wrong about CM1 concerns a large, circular erosion on the specimen's right (Broca erroneously says left) side of the frontal bone (figure 9.4). Broca thinks this feature lacks the characteristics of a pathological lesion, instead arguing it is postmortem damage from the skull having lain exposed on the cave floor. In a 2018 paper in *The Lancet*, Philippe Charlier and colleagues draw a different conclusion.[16] They think it *does* look like a disease process and generate CT and micro-CT images of the CM1 skull to reveal erosion of the outer table (scalp side) cortical bone in the absence of involvement of bone's internal (brain side) table. This rules out a malignant tumor; instead they believe the lesion to be a benign tumor known as neurofibromatosis type 1. They also find micro-CT evidence consistent with this diagnosis in CM1's ear canal.

What of the Balzi Rossi hominins? One feature that struck Verneau was how tall they were. Using long bone lengths to estimate their stature, five of the adult males appear to have been as tall or taller than 180 centimeters (5 feet 11 inches). Verneau estimates the stature of Barma Grande 1 at 188 centimeters (6 feet 2 inches), Grotte des Enfants 4 at 189 centimeters (6 feet 2.5 inches), and Barma

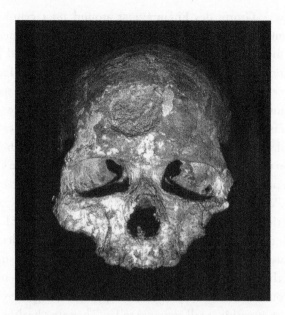

FIGURE 9.4 Anterior view of the Cro-Magnon 1 cranium. Note the large circular area of eroded bone on the forehead above the nose and right orbit.

Source: Courtesy of Erik Trinkaus.

Grande 2 at 177 centimeters (5 feet 10 inches). He is a bit more circumspect about the stature of the Baousso da Torre individuals, given the incomplete nature of their long bones, and his own doubts about Rivière's length estimates for these fragmentary specimens. He nonetheless concludes that BT1 was 185 centimeters (6 feet 1 inch) in height, and BT2 was about 180 centimeters (5 feet 11 inches). Even the females are tall; Verneau mistakenly thought Barma del Caviglione 1 was a male. He studied her skeleton in Paris and estimates her stature at about 175 centimeters (5 feet 9 inches). Although he considered the Grotte des Enfants 5 female to be a member of a different "race," he estimates her height at 159.5 centimeters (5 feet 3 inches), approximating the average height of French women at the beginning of the twentieth century. The adolescent male into whose burial she intrudes (Grotte des Enfants 6) had not yet reached his adult height; he was only 156 centimeters (5 feet 1.5 inches) tall. Verneau's stature estimates are almost certainly "in the ballpark" for these individuals.[17]

Employing stature estimation formulae widely used in forensic anthropology today, I calculate estimates for Barma Grande 1 at about 188.5 centimeters

(6 feet 2 inches), Grotte des Enfants 4 at around 181.4 centimeters (5 feet 11 inches), Barma Grande 2 at about 187 centimeters (6 feet 1.5 inches), Baousso da Torre 2 at about 178 centimeters (5 feet 10 inches), Barma del Caviglione at about 170 centimeters (5 feet 7 inches), and both Grotte des Enfants 5 and 6 at 158.2 centimeters (5 feet 2 inches). It is not solely the Balzi Rossi specimens who are tall; Trinkaus calculates the adult male Early Upper Paleolithic (n = 18) mean stature across Europe to be about 174.2 centimeters (5 feet 9 inches), with a minimum reported adult male stature of 165 centimeters (5 feet 5 inches).[18]

Some of you may be astonished at the estimated heights of the Cro-Magnons because this is not how we envision people in the past. I've spent a lot of time in Williamsburg, Virginia. My wife and I taught at William & Mary, and our son is now a student there, so I am quite familiar with the experience of having to duck through a door to get into a colonial-age building (Colonial Williamsburg is full of them). We are so used to thinking that people in the past were smaller than we are today that the stature estimates for the Balzi Rossi specimens may strike you as completely out of whack. They're not. European hunter-gatherers in the Late Pleistocene were every bit as tall as we are today. They ate a healthy diet that was rich in nutrients and particularly high in protein (and apparently vitamin D), and they lived at low enough population densities that epidemic and pandemic diseases were simply not an issue. As I write these words in the summer of 2020, this last point truly hits home.

Stature begins to decline globally in the Holocene, which began 11,700 years ago. The cause of this decline can be laid at the doorstep of agriculture. When people adopt agriculture, the detrimental health effects are immediately obvious. An agricultural diet is lower in protein, so agriculturalists tend not to grow as big as they otherwise would have. People must remain in one place to practice agriculture; you can't abandon your crops for a month or two and just hope for the best! In agricultural groups, population density ultimately goes up, thus increasing the likelihood of catching and transmitting a communicable disease. All of this conspires to keep body size low, and it remained low until the late nineteenth and early twentieth century when a "secular trend" takes off and stature rises again due to improved health care and better diets. This occurred first in Europe and North America and is now happening in Africa, Latin America, and South Asia.

The last three decades have also witnessed the rise of the field of bioarchaeology, which in North America is defined as the application of bioanthropological techniques to the recovery and analysis of human skeletons from archaeological contexts.[19] In this I discuss the bioarchaeology of three individuals interred together at the site of Dolní Věstonice II in Czechia—the so-called Triple Burial of Dolní Věstonice (DV) 13, 14, and 15. This 31,000-year-old burial of three young adults was discovered in August 1986 during a salvage operation led by Bohuslav Klíma and Jiří Svoboda (see chapter 2). According to Hillson et al., multiple lines of evidence suggest that DV 13 was about 21 to 24 years old at the time of death. DV 14 retains some unfused epiphyseal lines and incomplete roots of the wisdom teeth; he was probably about 17 to 19 at the time of death.[20] The pathological DV 15 falls between these two individuals in age (about 19 to 23). In terms of estimated stature, these individuals are not as tall as the tallest Balzi Rossi individuals, but apart from DV 15 they were not particularly short either. DV 13 was about 169 centimeters (5 feet 7 inches) tall; DV 14 was about 180 centimeters (5 feet 11 inches) tall; and the pathological DV 15 was about 160 centimeters (5 feet 3 inches) tall.

One thing I should mention about this "burial" is that it's not a burial per se. Rather, it appears that the three bodies were laid on the ground with their heads lower than their bodies due to the land's natural slope. It is possible that no burial pit was dug due to the difficulty of digging into permafrost. Humans then put a layer of glacial wind-blown soil (loess) over the bodies' red ochre–stained heads, and apparently to protect them from carnivores and the elements, the remainder of their bodies were covered with a structure made of sticks and other brush, some of which, branches of spruce, survive to this day. At some point this burial structure seems to have caught fire, which appears to have been quickly extinguished. In my own study of the skeletons, I found evidence of burning in DV 14's ninth, tenth, and eleventh thoracic vertebrae.

Figure 9.5 is a photograph of the three individuals during their recovery—from left to right they are DV 13, DV 15, and DV 14. Given the excellent state of preservation of the skeletons, each of which remains fully articulated, it appears the positioning of the three individuals, all lain in extended position side-by-side, was intentional. DV 13 and DV 15 are in the supine position, and DV 14 is laid face down—an unusual position we saw in chapter 1 with the Baousso da Torre child. DV 14 was the last of the three to be interred; his body lies atop DV 15's left arm. Also of interest is the fact that DV 13's arms are extended such that his hands are positioned on DV 15's crotch—an area with a heavy application of powdered red ochre. The pelves of DV 13 and 14 indicate male biological sex, but DV 15's host of pathological conditions (see below) led to some ambiguity regarding its biological sex. In 1994, I assessed his pelvis as male, and I have long wondered if at least some of those who thought the specimen was female were uncomfortable with the idea of one male's hands being positioned in another male's crotch. In 2016, this issue was resolved when Alissa Mittnik and colleagues extracted DNA from the skeletons and demonstrated that all three Triple Burial individuals are male.[21] These researchers also posit close maternal kinship between DV 14 and

FIGURE 9.5 The Dolní Věstonice Triple Burial as it appeared during excavation—left to right are DV 13, DV 15, and DV 14.

Source: Courtesy of Jiří Svoboda.

DV 15, which is reflected in their identical mtDNA; this could indicate that they were brothers! How these two brothers came to be buried side-by-side, one face down, with a third young adult male, in a structure that was partially burned, remains a mystery.

Dolní Věstonice 15 is in the center of the burial. His was the first body placed there, and some argue that he was central to any rituals associated with the triple burial. Some have suggested that people like DV 15 with unusual conditions were either venerated or treated differently in both life *and* death. Indeed, Trinkaus has shown that there is an unexpectedly high rate of developmental abnormalities among Cro-Magnon burials.[22] Given that high rates of developmental abnormalities are also observed in isolated, and therefore perhaps unburied, skeletal elements, he suspects the frequency of developmental abnormalities does not reflect selective funerary practices but rather was a biological reality for the Cro-Magnons. In light of the genetic homogeneity we see among Gravettian people across Europe (chapter 6), such developmental abnormalities *could* be due to inbreeding.

Dolní Věstonice 15 has a host of pathological conditions. The etiology of his disease(s) remain(s) uncertain, but what can be said is that he suffered from some sort of congenital condition that affected the morphology of his limb bones. This also made him shorter than most Cro-Magnons. His small size was likely in part due to systemic stress he experienced as a child that shows up in the form of enamel hypoplasias—areas of his adult teeth on which little to no enamel was laid down during their formation long before their eruption. DV 15 also shows abnormal curvature of the bones of his upper and lower limbs; his humeri and femora are the most curved.

Despite these pathologies, Trinkaus and colleagues argue the morphology of his hip and knee joints, as well as the pattern of cortical bone in his lower limbs,[23] suggest he was capable of walking in a normal manner, and that the curvature of his arms may have merely been the result of his trying to do heavy work.[24] In the end, despite his congenital abnormalities, DV 15 looks to have led an extremely active, albeit short life!

Dolní Věstonice 13's cranium shows three injuries that had healed long before his death—one on the frontal bone above the right orbit, one on the right parietal close to the frontal, and a raised area of bone on the temporal near the back of the skull. Whether these injuries were sustained simultaneously, or how they came about, is uncertain. In 1991, Vlček argued that two long front-to-back fractures of his parietal bones were likely related to a blow that killed him. Vlček also maintained that a break on the back of the DV 14 skull resulted in his death. However, analysis by Trinkaus et al. suggests it is more likely that these breaks in DV 13 and 14 occurred from the weight of sediments accruing on top of the crania in the millennia following their burial.[25] Here we are confronted with the old tongue-in-cheek bioarchaeological adage about people whose skeletons suggest they were in perfect health—aside from the fact that they're dead!

I'd like to finish this section on the Triple Burial with a personal story about my own experience studying them. When I first examined DV 13 in 1999, I noticed a furrow, or channel, on the front of the body of the sixth cervical (neck) vertebra (figure 9.6). It ran from the lower left corner of the vertebral body up toward midline. With smooth edges, it did not appear to be the result of peri- or postmortem damage. I could not figure out what had caused the development of such a sulcus. At first, I thought it was due to an unusual configuration of the vertebral arteries—two arteries that make their way up to the brain through foramina (holes) at the sides of the top six neck vertebrae. But this didn't make sense to me because in life the fronts of the vertebral bodies are protected by a thick band of connective tissue known as the anterior longitudinal ligament— how would blood coursing through a relatively small artery generate a deep bony furrow through such a thick ligament? At some point I was discussing this question with a student, and she asked to see a photo of the burial. Upon seeing the

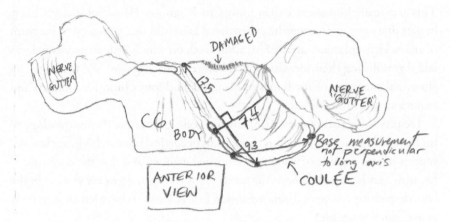

FIGURE 9.6 A drawing I made in 1999 of the Dolní Věstonice 13 sixth cervical vertebra. The furrow marked "coulée" (French for "gully") is the feature I found hard to explain.

Source: Drawing by author.

picture, she asked if it was perhaps due to the mandible resting on the vertebra in question. In figure 9.5, you can see that this looks to be the case! I thought hers was a brilliant observation, so the next time I went to Dolní Věstonice, I placed the bottom of the mandible into the furrow of C6, and it was a perfect fit. To this day, I remain baffled at how the bone was plastic enough to deform in such a smooth way that it mimicked an antemortem process.

BONES, BODIES, AND BEHAVIOR

Everyone knows the skeleton provides a protective framework for the body, but what many fail to realize is that bone is a dynamic, living tissue. For example, 25 percent of the blood's calcium is interchanged with bone calcium *every minute*! Bone also constantly responds to the biomechanical environment in which it finds itself. High activity levels generate forces many times one's own body weight, which affects bone. Force per unit area (stress) on bone causes *strain*, which is the deformation of bony tissue. In areas of a bone experiencing marked but intermittent strain, more bone is laid down so the bone is better able to resist strain in the future. This is particularly true of long bone shafts, which can grow in diameter in response to biomechanical stress even after growth in length of the long bone has ceased. As a result of this process, in most humans the long bones of the arm on the dominant side have thicker shafts than those on the nondominant side, a phenomenon known as directional asymmetry. About 90 percent of people are right-handed, so right arm bones tend to be more robustly built than those on the left. The bones of the lower limb are more symmetrical because the left and right legs make roughly equal contributions to locomotion. In contrast, our upper limbs are freed from locomotor constraints.

I finish this chapter with a discussion of the unusually asymmetrical arm bones of Barma Grande 2 (BG2). BG2 is one of the large male Balzi Rossi skeletons of likely Gravettian age you met in chapter 1. As evident in figure 9.7, his humeri look as if they belong to two different individuals, with the right (dominant) arm showing a much thicker and stronger shaft than the left. Yet the size of the humeral heads and articulations for the elbow joint are only a few millimeters different between the left and right side. Despite appearances, then, these humeri have not been mixed up in the lab; they belong to the same individual. What to make of this high left-right asymmetry? The answer is complicated.

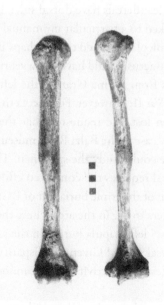

FIGURE 9.7 The Barma Grande 2 humeri in anterior view. The right humerus is on the left, and it has a much thicker shaft than the left humerus, which is on the right.

Source: Courtesy of Vincenzo Formicola.

First, starting in the 1980s, Trinkaus discovered that Neandertals were characterized by a high degree of upper limb cortical bone asymmetry, in which the bones of the dominant side, usually the right, were much more strongly built than the nondominant (usually left) ones. Over the years, researchers such as Steve Churchill hypothesized that this may have been related to the Neandertals' use of thrusting spears, in which the nondominant arm is used to guide the spear, and the power behind the thrust comes from the trailing, dominant arm. One could, however, imagine that other activities, such as throwing a spear or pulling a bowstring to launch an arrow, might lead to similar asymmetries between the left and right arms (chapter 10).

There's another Neandertal angle to this story as well. Vincenzo Formicola noted that the right scapula, or shoulder blade, of BG2 had a dorsal sulcus, or linear depression, along the side facing the axilla, or armpit. This anatomical feature is evident when a shallow depression associated with the attachment site for one of the rotator cuff muscles (teres minor) is found not on the ribcage side of the scapula, as it is in most modern humans, but rather is located on the scapula's dorsal, or back side. It is important to note that almost every Neandertal that preserves this anatomical region has a dorsal sulcus, making them different from recent humans, *Homo erectus*, or even "Lucy."

We don't know why Neandertals have dorsal sulci, but could it be related to high activity levels or linked to a particular unimanual (one-handed) behavior? If this were the case, Formicola reasoned that perhaps BG2's left scapula, associated with the "wimpy" humerus, would have had a ventral sulcus. Unfortunately, like many of the remains from Barma Grande, the left scapula was thought to have been lost in World War II. However, Formicola thought there was a chance the scapula had not been lost. He requested that the giant glass covering be removed from the display case at the Balzi Rossi museum (where BG2 is on permanent exhibition) so he could study the specimen. This glass cover is so large and heavy that its removal requires the concerted effort of several people. The display is a reconstruction of the triple burial pit of BG2, 3, and 4, and it is filled with sand. Using his fingers to dig in the area where the left shoulder would be, Formicola found the "lost" left scapula buried in the sand. Like its counterpart on the right, it had a dorsal sulcus! Given the disparity in size between the left and right arms, this suggests that activity alone cannot explain the presence or absence of a dorsal sulcus.

In 1997, Formicola teamed up with Churchill to analyze the asymmetry of BG2's arms. One finding of interest they made is that, while BG2's upper limb asymmetry is extreme and warrants explanation, Cro-Magnon males in general are characterized by greater left-right asymmetry in arm bone strength than are recent human males—even greater than some hunter-gatherers of the ethnographic present! Figure 9.8 is modified from their study and shows box-and-whiskers plots for cross-sectional cortical area asymmetry in three recent human male samples (Native Americans, Aleuts, and European Americans), the Cro-Magnons (divided into Early and Late Upper Paleolithic samples), and Neandertals.[26] You can see that the Cro-Magnon males have higher left-right asymmetry than any of the recent human samples. In fact, their median percentage asymmetry values (26.2 and 29.2) are higher than those of the Neandertals (16.5)! Among recent humans, the only hunter-gatherer group analyzed (the Aleuts) have the highest median asymmetry value (9.5), and the Native American and European American samples have the lowest median asymmetry at 5.4 and 6.0, respectively.

The black dots are fossil hominins characterized by pathology or disuse atrophy of one of their limbs, and for this reason neither they nor BG2 was included

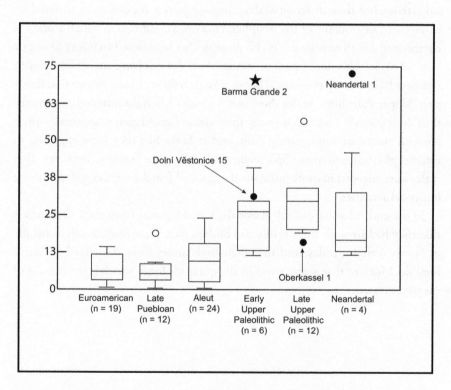

FIGURE 9.8 Box-and-whiskers plot of left-right asymmetry in humeral cortical bone cross-sectional area at the 35 percent section for male Neandertals, Cro-Magnons (Early and Late Upper Paleolithic), and recent humans (Euroamericans, Puebloans, and Aleuts).

Source: Modified and redrawn from Steven E. Churchill and Vincenzo Formicola, "A Case of Marked Bilateral Asymmetry in the Upper Limbs of an Upper Palaeolithic Male from Barma Grande (Liguria), Italy," *International Journal of Osteoarchaeology* 7 (1997): 18–38.

in the construction of the box plots. The Feldhofer Neandertal has a healed left ulnar fracture that deformed his elbow. This led to him preferentially using, and ultimately strengthening, his right arm, while his left arm suffered from disuse atrophy. As evident in figure 9.8, at 72.7 percent, the Feldhofer Neandertal has the highest degree of left-right cortical asymmetry of the entire sample. Two Cro-Magnon specimens, Dolní Věstonice 15 (DV15) and Oberkassel 1 (OBK1), also show pathological upper limbs. I have discussed DV 15's pathologies; OBK1 has a poorly healed fracture of his right ulna at its proximal, or elbow, end. Neither specimen's upper limb asymmetry is as extreme as that of the original Neandertal. Although there is no obvious injury that might have led to disuse atrophy, at 68.4 percent asymmetry Barma Grande 2 is second only to Neandertal 1 in its extreme right-left arm strength asymmetry.

Churchill and Formicola say they cannot rule out BG2 as nonpathological, and therefore his extreme asymmetry is the result of some unknown unimanual activity, but their differential diagnosis suggests it is possible he suffered an entrapment neuropathy of the peripheral nerves, a condition in which a nerve is compressed due to an injury. It is also possible that he suffered an injury to one or more of his left shoulder muscles—trauma that did not leave any skeletal traces. I suspect BG2 is nonpathological because Sparacello et al. have shown that European Upper Paleolithic males show much greater left-right humeral asymmetry than do females.[27] This is not to say that female Cro-Magnons are weak—their humeral shafts are impressively built, and it looks like they were engaging in intense physical activities. Sparacello and colleagues suspect, however, that males were engaged in more unimanual tasks, and females were engaged in more bimanual activities.

In the end, what we can say about the Cro-Magnons from their skeletons is that they had strongly built skulls and bodies, were on average as tall as people in North America today, and that their limb bones show they lived an active lifestyle. Much of that activity was in the quest for food, which is the subject of the next chapter.

10

Slings and Arrows

Straffa preferred his own company, which was a good thing because he was alone most of the time. Banished from his group due to a relationship gone sour, he had taken to life as a trader. There were not many who chose his way of life because it was a lonely existence, but he had gotten used to it and over the last decade had learned what commodities people wanted and how to get those goods to them. He made a regular circuit from the Adriatic Sea in what is now Montenegro to the broad plain north of the Sudeten Mountains; because he brought exotic goods for a fair price, he was warmly welcomed almost everywhere he went.

In one sense, he was never completely alone because he was always accompanied by his two wolf-dogs, Sima and Siffa. He had bought them as puppies at one of his regular stops in what is now the Bečva Valley of Czechia. The seller's one admonition was that under no circumstances was Straffa to feed them mammoth meat. Were they to eat it, they would revert to their wild nature and abandon him. He had followed that advice, and as a result his dogs were fiercely loyal, keeping him safe on his long and lonely circuit. The only downside to having them around was that at times they intimidated valued customers. The most helpful thing they did was that each pulled a travois loaded with trade goods. Straffa and his dogs brought seashells, wild figs, and pine nuts from the Adriatic up to the Middle Danube Valley, where they were highly prized. In the far north of his circuit, in what is now Poland, he acquired arctic fox pelts and black jet gemstones. These he brought south, where they were considered luxury items. He acquired obsidian whenever and wherever he could. This commodity varied wildly in its availability, and it fetched a steep price in areas where it was unavailable.

Straffa delighted in the bright and clear morning—what a glorious day to be alive! There was little wind, so he could hear the rhythmic beat of his steps, the panting of his dogs, and the sound of the travois sliding through the grass. The sweet smell of spruce wafted toward him. In the distance, Pálava Hill rose high above the plain, with a mix of conifers and grass adorning its sides. This was the second day he had been making his way toward the landmark, and today he would reach it. His first stop would be one of his favorites: a village at its base—he could see smoke rising from where he knew it to be. The people lived there year-round, which meant he did not have to worry about having missed them. They were mammoth hunters, and they *always* had a surplus of meat. He could acquire all the meat he would need for the entire year at this one spot. But the prize he really hoped to secure here was ivory, which the people here had in excess. But they knew its value, so it wouldn't come cheap. He hoped their desire for seashells, jet, and obsidian was high this year.

Within an hour he saw the mammoth-bone huts of the village and caught sight of Mora and Bilan raising their hands in greeting. Aside from the dogs, these two were the closest thing he had to friends these days. He found himself looking forward to supping on mammoth with them this evening; it had been too long since that rich meat had crossed his lips. As he returned their greeting, he began to salivate in anticipation of the night ahead. Then he quickly remembered he had to ensure that his dogs were not given any scraps.

In the first half of this chapter, I discuss how the Cro-Magnons acquired the animals they ate. In chapter 7, I alluded to the importance of long-distance projectile technology for the success of early modern humans; here I look at those data in a more fine-grained way. In the second half of the chapter, I explain what we know about the Cro-Magnon diet (both its animal and plant components), taking advantage of recent advances in archaeology, including stable isotope analyses.

THE EARLIEST PROJECTILES?

How old is projectile technology? In 1997 Harmut Thieme reported an extraordinary find of beautifully preserved organic tools from 330,000-year-old Acheulean contexts in Schöningen, Germany. Perhaps most amazing among these were three wooden spears and a potential throwing stick, which may have been thrown like boomerangs are today. With these items were stone tools and the butchered carcasses of more than ten horses. Thieme reported the discovery of four more spears from Schöningen in 1999. The spears are all around 2 meters (6.5 feet) in length, and each is made from the trunk of a spruce sapling (genus *Picea*) from which the limbs and bark have been removed. The spears are sharpened to a point but stone tips are not hafted onto them. The base of the trunk served as the distal or "business end" of the spear. As a result, the point end of the spear is heaviest (and strongest), with its center of gravity displaced distally on the shaft. Thieme compares these spears to front-heavy modern javelins, and he maintains that the Schöningen spears are projectile weapons.[1] If they are projectiles, along with the distal end of an even older (about 450 ka BP) yew spear (genus *Taxus*) found in 1911 in Clacton, England, they would be the oldest projectiles yet found.

We do not know which species manufactured these spears, but given their age the Schöningen spears were most likely made by early Neandertals. The older Clacton spear could antedate the Neandertals, in which case *H. heidelbergensis* is a possibility. There is, however, every reason to believe that Neandertals continued to use this technology well into the Late Pleistocene, which began 129,000 years ago. Indeed, a similarly sharpened spear made from yew like the Clacton spear dates to about 125 ka BP. It was recovered in 1948 at another site in Germany called Lehringen. The Lehringen spear was found within the rib cage of a straight-tusked elephant (genus *Palaeoloxodon*), suggesting to many that it was used in the hunting of large-bodied prey.

Thieme maintains that tests of replicas of the Schöningen spears highlight their aerodynamic properties and are consistent with their use as projectiles. However, John Shea and Steven Churchill find this claim dubious.[2] Both note that the thickness of the spears from all three sites is much greater than those of modern throwing spears and more similar to, and on average even thicker than, the diameter of ethnographic examples of so-called thrusting spears. Churchill also reports experiments with students, some of whom were experienced javelin throwers, in which replicas of the Schöningen spears performed poorly in hitting a hay bale target at a distance of >15 meters (about 50 feet)—"more often than not, the spears tumbled in flight, ending up striking the target sideways or butt-first, or missing the target altogether."[3]

Churchill is quick to point out that this does not mean the Schöningen spears were never thrown; it's just that in the ethnographic present handheld spears like this are close-range weapons, usually employed against large-sized prey at a distance of 6 meters (20 feet) or closer. I am almost certainly less brave than my Neandertal ancestors, but one reason not to throw your thrusting spear is that if you miss the target or fail to hit a vital area you now have an angry animal that views you as a threat. And guess what? You no longer have a spear! In the ethnographic present, Churchill also says that thrusting spears are primarily used to take larger prey during disadvantage hunting, in which game are first driven into a natural trap.[4] This method usually involves multiple hunters. If you happen to lose your spear, you hope one of your friends "has your back" and that the trapped animal is immobilized as well.

You may be wondering why these earliest spears are not tipped with stone points. Churchill notes that lithic points cut better and tend to cause more internal damage to prey than do sharpened organic points (this has been demonstrated in experimental studies with gelatin targets), but stone points are much more brittle than organic ones and are more frequently subject to breakage.[5] Thus there seems to be a trade-off between a spear's durability and its wound effectiveness.

If these earliest spears are thrusting spears, or very short-range throwing spears, when do long-range projectile weapons emerge? We must keep in mind that simple slings for hurling stones a great distance are easily made from hides or sinew. Unfortunately, such weapons are archaeologically invisible. To find the earliest archaeological evidence for long-range projectiles, Shea examined lithic points from Middle and Upper Paleolithic contexts in Africa, the Levant, and Europe and compared them to examples of (from smallest to largest) arrowheads, dart tips, and experimentally made thrusting spear tips.[6] He measured the bases of stone points rather than the tips because all too frequently the tips of stone points have been broken off, in many cases because they were used as projectiles! He employs an approximation of the point's base area known as the tip cross-sectional area (TCSA), which is calculated this way:[7]

(0.5 × maximum width of base in mm) × maximum thickness of base in mm

Unless the point has a smaller extension or tang at its base to facilitate hafting, the shaft of a spear or arrow is the same width or wider than the base of the stone point hafted onto it. What does Shea find? First, in Europe, Neandertal-made Middle Paleolithic, Mousterian and Levallois points from Le Moustier, as well as Middle Paleolithic foliate (leaf-shaped) points from Mauern, Germany, are all significantly larger than ethnographic examples of projectile points (figure 10.1). If they were spear points, this suggests that they were hafted onto thrusting spears. In the Levant, where both modern humans and Neandertals are associated with the Middle Paleolithic, Shea found similar patterns. No matter which site or which species is presumed to be making them, the Levantine Middle Paleolithic points

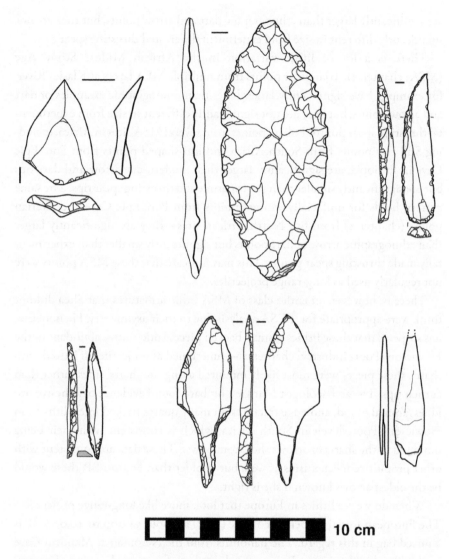

10 cm

FIGURE 10.1 Potential Paleolithic projectiles drawn to the same scale. Clockwise from upper left, three views of a Levallois point from Kebara, Israel; two views of a Middle Paleolithic foliate point from Mauern, Germany; three views of a Châtelperronian point from Les Cottés, France; two views of an antler Aurignacian split-base point from El Castillo, Spain; three views of a Font Robert point from the site of Font Robert, France; and three views of a Gravette point from La Gravette, France.

Source: The Aurignacian point is redrawn from José-Miguel Tejero, "Spanish Aurignacian Projectile Points: An Example of the First European Paleolithic Hunting Weapons in Osseous Materials," in *Osseous Projectile Weaponry: Towards an Understanding of Pleistocene Cultural Variability*, ed. Michelle C. Langley (Dordrecht: Springer, 2016), 55–69. All other images and scale are from John J. Shea, *Stone Tools in Human Evolution: Behavioral Differences Among Technological Primates* (Cambridge: Cambridge University Press, 2017), courtesy of John J. Shea.

are significantly larger than ethnographic dart and arrow points, but they are not significantly different in size from experimentally created thrusting spear tips.

There is a lot of lithic variation in the African Middle Stone Age (MSA; chapter 7). Triangular points from multiple MSA layers at Klasies River, for example, have significantly larger bases than ethnographic examples of dart and arrow points, but they are not significantly different in size from experimental thrusting spear points. Foliate points from several MSA sites in Africa, including Still Bay points from South Africa and leaf-shaped points from Porc-Epic Cave in Ethiopia, are significantly larger than modern-day arrow and dart tips but are significantly smaller than the experimental thrusting spear tips. The same pattern holds for unifacially worked points from Porc-Epic Cave and Aterian points (chapter 2) from five North African sites—they are significantly larger than ethnographic arrow or dart points but significantly smaller than experimentally made thrusting spear points. This may indicate that these MSA points were not regularly used as long-range projectiles.

There is, however, an entire class of MSA lithic armatures that Shea did not think were appropriate for TCSA analysis due to their asymmetry. He nonetheless suspects that these lithics—small backed pieces (microliths, abundant in the Howiesons Poort industry; chapter 7)—functioned as projectiles. If hafted onto shafts, these pieces were most likely mounted along the shafts' sides rather than at their tips. The archaeologist Marlize Lombard found evidence of adhesive residues, animal blood, and edge fractures on many quartz backed microliths from Howiesons Poort levels in South Africa, which is consistent with their being mounted in the shafts of arrows shot from bows.[8] These data are consistent with other propulsive mechanisms as well, but at older than 60,000 BP, these would be the oldest arrows known if she is right.

When do we see lithics in Europe that look more like long-range projectiles? The European Initial Upper Paleolithic (IUP) at about 54,000 to 40,000 BP is a mixed bag in this regard. The nanoliths from the Neronian at Mandrin Cave could, like the Howiesons Poort microliths, have functioned as part of projectiles. Similarly, Shea suspects that Uluzzian lunate pieces may have functioned in much the same way as the backed pieces in the MSA—in other words, they were mounted along the shafts of projectile weapons rather than at their tips.[9] There is now evidence to back Shea's hypothesis. In 2019, Sano and colleagues reported that many of the lunate bladelets from the Uluzzian type site of Cavallo in southern Italy showed diagnostic impact fractures consistent with their having been used as high velocity projectiles. Residues from the backed portion of the lithics show an adhesive made from ochre, tree gum, and beeswax was used to affix them to a shaft, and many of the lithics' breakage patterns suggest that they were mounted at the tip rather than along the side of the shaft. Sano et al.'s TCSA data for these bladelets show that many of them are even smaller than ethnographic examples of arrowheads; this could be the earliest evidence for the bow in Europe![10]

In contrast, Shea found that Central European Szeletian foliate points are significantly larger than the ethnographic dart and arrow points, but not significantly different in size from the experimentally created thrusting spear points. If they were mounted onto spears, this suggests that they were of the thrusting spear variety. Also in Central Europe, Bohunician foliate points are significantly bigger than ethnographic spear and arrow tips and are significantly larger than the experimental thrusting spear tips as well!

Finally, although Châtelperron points (see figure 10.1) are significantly larger than ethnographic examples of arrowheads, they are not significantly bigger than ethnographic dart points. Shea notes, too, that breakage patterns of many Châtelperron points are consistent with their having been used as projectiles. It is therefore plausible that at least some Châtelperron points were launched as projectiles. We thus have archaeological evidence for the use of long-distance projectile weapons for some, but not all, IUP industries.

We see an explosion in long-range projectile weapon technology in the Aurignacian about 43,000 years ago. Over the past fifteen years there has been a growing appreciation for the fact that many Aurignacian tools, especially artifacts previously called scrapers and burins, were not tools per se but rather were cores from which small bladelets were repeatedly struck. Bladelets are a major component of the Aurignacian (and Protoaurignacian) tool kit. As such, these bladelets could have functioned much in the way the triangular Howiesons Poort pieces in South Africa or Uluzzian lunates are suspected of being used; that is, they were embedded along the shafts, if not the tips, of projectile weapons.

Shea does not report any measurements for Aurignacian lithic points, but keep in mind that the Aurignacian, and especially the early Aurignacian, tends to be characterized by a high proportion of antler, bone, and even ivory points. In fact, for well over a hundred years, Aurignacian split-based points, most frequently made from antler, have nearly universally been viewed by archaeologists as projectile weapons (see figure 10.1). In this light, I was able to cull width and thickness measurements for fifty-four Aurignacian split-based antler points from Tartar and White.[11] These artifacts come from two sites (Abris Blanchard and Castanet) in the Castel-Merle complex in France, located just 10 kilometers (6.2 miles) NE of Les Eyzies. I calculate the mean TCSA for these split-based antler points at 46.4 millimeters^2, with a standard deviation of 20.5. This is statistically significantly larger than those of ethnographic arrowheads but statistically significantly *smaller* than both the ethnographic dart sample and the thrusting spear point sample. These data strongly suggest that Aurignacian split-based points are part of a long-range projectile weapon system.

Overlapping in time with and ultimately replacing the Aurignacian is the Gravettian, which first appears about 35,000 years ago. In the Gravettian, stone points are more frequently encountered than in the Aurignacian; in particular, Gravette points and tanged Font Robert points (see figure 10.1) dominate Gravettian assemblages. Shea found that both of these point types are significantly larger than ethnographic arrowheads. Although the Font Robert points are not significantly different in size from ethnographic dart tips, Gravette points (like the Aurignacian split-based points) are significantly *smaller* than dart tips. Gravette and Font Robert points thus appear to be part of projectile weapons as well.

Taken as a whole, these data suggest that long-range projectile weaponry only arrives in Europe with the Upper Paleolithic. Its presence in the IUP industries is spotty, but its use is universal by the Aurignacian and Gravettian. Remembering Churchill's ethnographic data, we know that sometimes these weapons were launched by hand, especially at close range. However, most archaeologists posit that these projectiles were much more frequently used with propulsive systems, especially in light of the high-impact breakage evident on many of them. The simplest propulsive technology is to launch spears or "darts" with the aid of a spear thrower. In North America, the spear thrower is commonly known by its Nahuatl, or Aztec, name, *atlatl*. The earliest recovered atlatls were made from antler and postdate the Aurignacian or Gravettian, coming from Solutrean contexts (about 24–18 ka BP). We have every reason to believe that antler atlatls were preceded for many millennia by wooden ones—tools not preserved in the archaeological record.

The atlatl is a simple beam about 30–60 centimeters (12–24 inches) long. At one end is a grip, which may include a strap made of sinew or leather; its other end features a small protuberance called the spur. The spur fits into a hollow carved into the base of a long (about 150–210 centimeters [5–7 feet]) lightweight dart. The dart itself is usually fletched with feathers or leaves. The atlatl acts as an extension of the thrower's arm, providing extra leverage for launching the spear with more power, speed, and at a greater distance. It is a simple yet highly effective technology. Churchill notes that a dart thrown with the aid of an atlatl has an effective range of about 39.6 meters (130 feet), making it a truly long-range weapon. By way of contrast, a hand-thrown spear has an effective range of only 7.8 meters (26 feet).[12]

What about the bow and arrow? There are indications that the bow was used in Africa more than 60,000 years ago, and Sano et al. say one cannot exclude the possibility that the Uluzzian lunate pieces from Cavallo were arrowheads.[13] There is no reason to think that bows and arrows were an unknown technology in Europe about 45,000 years ago. Despite this fact, it is unlikely that most of the European Upper Paleolithic points were arrowheads for a variety of reasons. First, although the archaeologist Michelle Langley shows there is considerable size overlap between darts propelled by spear throwers and arrows launched from a bow, the split-based Aurignacian points and Gravettian points are nonetheless statistically significantly bigger than ethnographic examples of lithic arrowheads.[14] Second, the bow and arrow is a versatile technology that can be used effectively against a host of prey animals, but when killing larger animals, the generally smaller arrow must hit vital organs within the thoracic cavity and avoid the ribs and scapulae (shoulder blades). The space between ribs in even an enormous mammal like a mammoth is a relatively small target area, and as such, the effective range of the bow and arrow is much shorter (25.8 meters [85 feet]) than that of the atlatl dart.[15] Finally, the shorter effective range of the bow and arrow makes it less ideal than the atlatl for hunting game in an open environment. Out in the open it is harder to conceal oneself to get within range of the prey, and much of Europe in the Pleistocene was treeless steppe/tundra. In a densely forested ecosystem, however, the more "surgical" nature of the bow and arrow, in an environment in which ambush or encounter hunting can more easily lead to close encounters between hunter and prey, likely favors its use over the atlatl. As the Pleistocene drew to a close about 11,700 years ago, Europe became more forested, leading to the extinction of many of the grass- and tundra-adapted megafauna that had long dominated the continent (chapter 12). Given smaller overall prey size and an increasingly more forested environment, hunters almost certainly increased their reliance on the bow and arrow. This technological shift is evident in the greater focus on the manufacture of microliths during the Epipaleolithic and Mesolithic of the early Holocene—but better still, we *know* people employed the bow during the Mesolithic because its use is depicted in about 7,000-year-old Mesolithic rock art in the Alta Fjord in northern Norway!

In chapter 7 I mentioned what a significant innovation long-range projectile weaponry was. I cannot stress enough just how important this technology was to our ancestors. The ability to kill animals, especially dangerous ones, from a distance would have had huge effects on group safety, and it would rapidly bring about increases in survivorship and fertility in the groups employing these technologies. Perhaps it is no surprise then that the earliest evidence we have for a significant increase in population size in Europe is associated with the Aurignacian. This first population boom was followed by what looks to be an even more significant increase in population size associated with the Gravettian—a time in which common art motifs flourished across the continent.

WHAT'S FOR DINNER? THE CRO-MAGNON DIET

What were the Cro-Magnons eating? Faunal remains tend to be overrepresented in the archaeological record when compared to perishable resources such as plant residues, and archaeologists have long focused on meat consumption, in particular the consumption of large mammals. They study faunal remains accumulated by ancient hunters, especially those bones showing cut marks or burning, because they know these remains are present due to human activity. Their focus on meat may not be problematic, however. Cro-Magnons lived at high latitudes and would have been more reliant on animal than plant foods; this would have been further intensified during colder stadials. This is not to say that plants were not consumed—plants are a great source of calories and provide nonenergetic nutrients that may be absent in meat. One such nutrient is vitamin C. Unlike most carnivores, humans cannot synthesize vitamin C. We must consume it in food, and plants are where one tends to find it.

What does the zooarchaeology of the Cro-Magnons tell us? For the duration of the Upper Paleolithic, those who lived along the Mediterranean and Atlantic coasts exploited shellfish and other marine resources. Their primary subsistence, however, remained focused on terrestrial animals. In Europe south of the Alps, the main terrestrial game taken by the Cro-Magnons were red deer (*Cervus elaphus*, the same genus as North American elk), wild goats called ibex (*Capra ibex*), and the wild ancestors of domestic cattle known as aurochs (*Bos primigenius*). In Europe north of the Alps, the primary species taken were reindeer (*Rangifer tarandus*), but horse (*Equus ferus*) and red deer were also hunted. Many reindeer are migratory and, depending on food availability, the reindeer of southwestern France may have been as well. It is possible they spent winters in the sheltered lowlands, dispersed in riverine forests in groups as small as a mother with her dependent offspring. With the return of spring, they may have become gregarious, forming huge herds that migrated to feed and calve in the higher elevation grasslands. These herds may have headed south to the Pyrenees or east, following the Dordogne, Vézère, and other river valleys to the Massif Central. Many French Upper Paleolithic sites were places where humans were almost singularly focused on the acquisition of reindeer. At the Abri Pataud, for example, throughout the Upper Paleolithic sequence, somewhere between 80 and 90 percent of the identifiable mammalian bones are those of reindeer. They also appear to have been hunted in winter. The aptly named Grotte du Renne (Reindeer Cave) at Arcy-sur-Cure in north-central France has a similar faunal signature.

Britain was connected to the continent during the Late Pleistocene. What is now the English Channel was a broad, lush valley with a massive river fed by the Thames, Seine, and Rhine. Here, too, we imagine large herds of reindeer congregating in their summer feeding and calving grounds—grounds now beneath the sea. If the Cro-Magnons hunting these herds could position themselves at strategic chokepoints, they could take advantage of the hundreds of thousands of migrating reindeer as they moved from one ecozone to another in both the spring and fall migrations. The latter migration would be the more important one because the animals would have accumulated body fat for the winter. Many of the sites inhabited by these hunters remain beneath the sea, but we do have evidence for the exploitation of reindeer in the Upper Paleolithic levels at British sites such as Paviland and Gough's Cave.

Lewis Binford used to tell a story that almost certainly has parallels to our Upper Paleolithic reindeer hunters. From 1969 to 1974 he did ethnoarchaeology among the Nunamiut, Inuit Iñupiaq speakers who live in Anaktuvuk Pass in Alaska's Brooks Range.[16] The Nunamiut are hunters of reindeer (caribou) whose mountain pass home is a choke point for their annual migrations. The caribou the Nunamiut hunt spend winter in forests south of the Brooks Range, either individually or in small groups. But they congregate in huge numbers in the spring and spend their summers north of the mountains, giving birth and gorging themselves on the tundra of Alaska's North Slope. In the fall, perhaps spurred on by the first snowfall, they return to cross south over the mountains en masse. The fall hunt is the most critical hunt of the year for the Nunamiut. Binford himself observed many caribou hunts and heard village elders recount stories of fabled hunts from long ago. For me, however, the most poignant story Lew told about the Nunamiut was one concerning a young boy and his grandfather.

Binford says he happened to come across the old man and boy at the edge of a pond. There was a float made from a caribou stomach out in the water, and the man was teaching his grandson how to throw a harpoon at it. Lew said he had never seen anyone doing anything even remotely similar to this, and with curiosity finally getting the better of him he asked what was going on. The grandfather said he was teaching his son how to hunt seals in case the caribou fail to come. If the caribou don't come, he told Binford, we will have to move to the coast to hunt seals. Keep in mind that Anaktuvuk Pass is about 250 kilometers (156 miles) from the closest coast.

Binford later investigated and discovered that about every 120 years or so massive forest fires south of the Brooks Range lead to a crash in the caribou population. So we have a grandfather, who may have never seen a seal in his life, teaching his grandson how to hunt seals, just as his grandfather had taught him, in case the caribou crash happens during his grandson's lifetime and the Nunamiut are forced to move to the coast to take up seal hunting. This is the sort of intergenerational contingency planning that goes hand-in-hand with modern human hunting and gathering, and I am certain that our Upper Paleolithic hunters had similar strategies for dealing with these sorts of dire situations (much like Grandma Ela's group did in the preface).

MAMMOTH UNDERTAKINGS

During the Pleistocene, a vast grassland stretched from the middle of the Danube Valley in what is now Austria, Czechia, and Slovakia, downstream to the Black Sea (at the time a lake), and then for thousands of kilometers east across what is now Romania, Moldova, Ukraine, Russia, and Kazakhstan. This grassland supported huge herds of large-bodied herbivorous mammals. Because of their large size, scientists call these mammals megafauna. The largesse of these animals provided food and raw material resources for Paleolithic hunter-gatherers for more than 30,000 years. Like their counterparts in western Europe, hunters on the central and eastern European Plain took reindeer, red deer, horse, and bison; but on the seemingly endless steppe, their primary focus was on a much bigger quarry—woolly mammoths (*Mammuthus primigenius*). In this largely treeless environment devoid of rock shelters and caves, the bones of these animals served as one of the main construction materials for building the circular huts found at sites across this vast region. Mammoth bone and ivory were favored for tools and art objects, and mammoth scapulae were used to cover the burials of the Dolní Věstonice 3 and Pavlov 1 individuals. Mammoths simply dominate the faunal assemblages at many of these sites—at Moravian sites such as Předmostí and Dolní Věstonice I and II, whether measured by the number of individual specimens (NISP) or by the minimum number of individuals (MNI), mammoths are the predominant faunal species—and at Kraków Spadzista in Poland, mammoths make up 99 percent of the NISP!

You are already familiar with Dolní Věstonice. The first sites there were discovered due to the presence of a local brickyard for which loess was excavated for use in the brickmaking process. After World War II, many of the surrounding hillsides were terraced for vineyards, leading to the discovery of further sites in both Dolní Věstonice and the neighboring village of Pavlov. Finally, in the 1980s, the damming of the Dyje River to create a shallow reservoir required the extraction of yet more loess, leading to the discovery of even more sites.

Dolní Věstonice I is one of the larger sites, representing a big settlement located on high ground just above what would have been marshy terrain. It features a massive deposit of thousands of mammoth bones and multiple circular dwellings 5 to 6 meters (16 to 20 feet) in diameter, all with internal fireplaces. The roofs of these teepeelike structures probably consisted of a wooden framework packed internally with loess for insulation and draped with skins. The base of the structures were made of mammoth bones.

The most spectacular examples of mammoth-bone huts, narrowly dated to 18,500–17,000 cal BP, come from the middle and upper Dniepr River basin in what is now Ukraine. Two sites stand out in particular: Mejiriche (Mezhirich), in the Dniepr River Valley about 112 kilometers (70 miles) southeast of Kyiv; and Mezine, in the Desna River Valley some 240 kilometers (150 miles) northeast of Kyiv. These circular-to-ovoid dwellings are 4 to 6 meters (13 to 19 feet) in diameter. Their bases are made from mammoth crania, and the outside walls are frequently made from interlocking mammoth mandibles, as well as scapulae, pelves, and long bones such as femora, humeri, tibiae, fibulae, radii, and ulnae. As at Dolní Věstonice, these walls were sealed internally with loess, and the exterior was probably covered with animal skins. It also seems likely that mammoth tusks, as well as saplings, were used to form the roof. The interior walls of these huts were decorated with geometric designs, which is evident from the painted mammoth bones in Hut No. 1 at Mezine.[17]

Mammoths predominate across the sites dating to this period in the Dniepr Basin, but reindeer, bison, muskoxen, and woolly rhinoceros are also found, as well as arctic and red foxes, lynx, hares, and marmots (almost certainly exploited for their pelts). Returning to Czechia, mammoths dominate the fauna at Dolní Věstonice I and II, but the same cannot be said of the neighboring Pavlov I site. Although mammoth was still the predominant species used to make tools here, it is not as well represented among the fauna, which feature a greater number of reindeer and smaller animals such as birds, hares, and foxes. Horse and bison are also frequent; in contrast, few red deer or ibex were found. This suggests that the area around Pálava Hill was visited by a wide variety of species.

About 87 kilometers (54 miles) northeast of Dolní Věstonice lies the Czech site of Předmostí, which is best known for the remains of fifteen nearly complete human skeletons. Unfortunately most of these were destroyed in the waning days of World War II. However, Předmostí also has a rich faunal record, including an MNI of more than 1,000 mammoths, which dominate its faunal assemblage. Perhaps the most interesting discovery at Předmostí, however, were the remains of what Germonpré and colleagues refer to as Paleolithic dogs.[18] Dated at about 31,000 years ago, they are the earliest domesticated animals ever recovered from an archaeological site. The skulls of these animals, intentionally interred, one with a bone placed in its mouth, differ in shape from those of wolves in much the same way domestic dogs do. DNA extracted from these animals, however, shows that they are not ancestral to any living dogs. These ancient dogs, and similar remains from Goyet Cave in Belgium, suggest that dogs were domesticated (or semidomesticated) multiple times in prehistory. Pat Shipman even argues that these and similar wolf-dogs across Europe gave a hunting advantage to the Cro-Magnons, helping them outcompete the Neandertals.[19] Bocherens et al. dispute this interpretation.[20] Using isotopic analyses, they show that the Předmostí dogs ate a different diet than the Předmostí humans, which means that the dogs did not have primary access to mammoth kills. Bocherens and colleagues maintain that the wolf-dogs were used to aid with transport (i.e., pulling sleds or travois) not for hunting.

YOU ARE WHAT YOU EAT

The old adage that "you are what you eat" is in fact true at the molecular level. In chapter 2, I discussed radioactive isotopes of elements, which decay over time and are therefore particularly useful in radiometric dating. In contrast, stable isotopes—nonradioactive heavier or lighter versions of a particular element that occur in nature and do not decay (i.e., remain in their heavier or lighter state)— have shown their value in prehistoric diet reconstruction. One example is ^{13}C, a heavier stable carbon isotope that has proven useful due to different photosynthetic pathways taken by plants. Most temperate plants, including trees, shrubs, and grasses, tend to first incorporate atmospheric carbon dioxide into three-carbon molecules and are therefore referred to as C3 plants. Tropical grasses first incorporate CO_2 into four-carbon molecules and are known as C4 plants. C3 plants are more selective against the incorporation of ^{13}C into their tissues than are C4 plants, and thus they have lower $^{13}C : ^{12}C$ ratios. These ratios remain stable at higher trophic levels (i.e., as one ascends the food chain), so measuring this ratio in an animal's skeleton can tell you whether that animal ate C4 plants or had eaten animals that themselves had eaten C4 plants. In North America this technique has proven useful in showing when populations adopted intensive maize agriculture (maize is a C4 plant). There were no C4 plants in Ice Age Europe, so why look at ^{13}C: ^{12}C ratios? First, marine organisms have more ^{13}C in their tissues than do terrestrial organisms, so the frequent exploitation of marine plants and animals by the Cro-Magnons should be detectable. Organisms from freshwater rivers, lakes, and streams tend to have *lower* ^{13}C than terrestrial organisms because the carbon in freshwater ecosystems not only comes from the atmosphere but also from geological sources (erosion). Thus one should be able to use ^{13}C: ^{12}C ratios to detect the frequent exploitation of freshwater resources by humans in the Pleistocene.

Nitrogen isotopes in Pleistocene skeletons are also informative about ancient diet. More than 99 percent of nitrogen in nature is ^{14}N, but there is a heavier stable isotope of nitrogen (^{15}N) that is used in dietary reconstruction. Nitrogenous compounds are absorbed through plants' roots, but unlike the ^{13}C: ^{12}C ratio, the ^{15}N: ^{14}N ratio tends to change across trophic levels, with folivores showing lower ^{15}N: ^{14}N ratios and carnivores having greater amounts of ^{15}N in their tissues. Nitrogen signatures are also affected by the amount of marine plants and animals incorporated into the diet. Aquatic food webs tend to have more trophic levels than terrestrial ones; so marine carnivores and terrestrial animals that eat marine resources have higher ^{15}N in their tissues than those focusing on terrestrial resources (freshwater fish also have high ^{15}N levels). In recent years, researchers have turned to rare sulfur (^{34}S) and calcium (^{42}Ca and ^{44}Ca) stable isotopes to reveal dietary aspects as well.

The ratios of carbon and nitrogen isotopes have elucidated much about the Cro-Magnons' diet. More than twenty years ago, work by Michael Richards and colleagues revealed that, like Neandertals, Cro-Magnon skeletons about 30,000 years ago from sites in Russia, Czechia, and the United Kingdom tend to show elevated ^{15}N levels, which would be expected for a top predator. However, unlike the Neandertals, the Cro-Magnons also show relatively low ^{13}C levels. Richards and colleagues interpret these data as reflecting an increase in the exploitation of resources from freshwater ecosystems (fish, waterfowl) among the Cro-Magnons.[21]

In recent years these results have been questioned. First, in 2014 Hervé Bocherens and colleagues, working at the site of Kudaro Cave in the Republic of Georgia, noted that salmon bones are present throughout the Mousterian (presumed Neandertal) layers.[22] The problem is that Neandertals were not the only large carnivores to use the cave. Cave lions (*Panthera spelaea*) and the largely vegetarian cave bears (*Ursus spelaeus*) are also present throughout the layers;

therefore, they could be responsible for the accumulation of these fish. However, the amount of ^{34}S present in salmon is extremely elevated; neither cave lion nor cave bear remains from Kudaro shows elevated levels of this rare sulfur isotope, leading to the conclusion that the Neandertals must have been responsible for the salmon at the site. Thus, at least some Neandertals appear to have been exploiting aquatic resources.

Second, in 2019 Christoph Wissing and colleagues extracted isotopes from both Neandertal and Cro-Magnon skeletons from the site of Goyet Cave and from Neandertals from the nearby site of Spy.[23] Analyzing ^{13}C, ^{15}N, and ^{34}S levels, they were unable to distinguish the Neandertals from the Cro-Magnons who were their penecontemporaries in the same region. Although modern humans may have been better resource extractors than Neandertals, their conclusion was that these two top predators were exploiting many of the same resources!

In 2015, Hervé Bocherens and colleagues not only extracted isotopes from the Předmostí fauna, they also extracted some from the Předmostí 26 human mandible, one of the few Předmostí specimens to survive the Second World War.[24] They first looked at patterning among the wide variety of fossil herbivores from the site. They found that mammoths had far and away the greatest amount of ^{15}N in their bones of any of the plant-eating species. They attribute this to the fact that mammoths are obligate grazers, eating nothing but grass, therefore limiting them to the steppe. In contrast, many herbivores can also subsist on lichen—a resource available in tundra environments. Those animals that ate the most lichen also had the highest levels of ^{13}C in their tissues. For example, the muskoxen, and to a lesser extent reindeer, had low ^{15}N in the context of high ^{13}C levels—the opposite is true for mammoths; horses, woolly rhinoceroses, bison, and red deer fall in the middle. Předmostí 26 had the highest ^{15}N levels of the entire sample, which is consistent with a diet rich in mammoth. Pleistocene wolves were a close second to the human skeleton, suggesting that they also feasted on mammoths. In contrast, the isotopic signatures of cave lions, brown bear, wolverines, and the Paleolithic "wolf-dogs" mentioned previously suggest a focus on different prey, perhaps reindeer and muskoxen.

García-González and colleagues found that the skeleton of the 18.9–18.7 ka cal BP "Red Lady" of El Mirón Cave in northern Spain had such a high ^{15}N content that it was almost certain she had eaten marine resources.[25] This further bolsters the hypothesis that the presence of abundant salmon bones indicate exploitation of the salmon runs at that site (although most of her protein was probably still derived from red deer and ibex). In contrast, the penecontemporary skeleton from the French site of St.-Germaine-la-Rivière shows a much more terrestrial ^{15}N profile. Perhaps this is not surprising because it is farther inland than El Mirón, and due to Pleistocene ice caps it would have been farther inland still at that time, given the broad continental shelf off the coast of southwestern France.

In 2021 I was fortunate to be involved in a research project led by Pierre-Jean Dodat, who is pioneering the use of rare isotopes of calcium (^{42}Ca and ^{44}Ca) in dietary reconstruction.[26] We were interested in using calcium isotopes to understand the diet of the Regourdou 1 Neandertal that at between 80,000 and 91,000 years old does not preserve sufficient amounts of carbon or nitrogen isotopes for analysis. However, the calcium salts in bone remain available nearly indefinitely. What we noted is that herbivores tend to have higher ^{44}Ca: ^{42}Ca ratios than carnivores, and Regourdou fell among carnivores in that regard. What is interesting is that Regourdou's high ratio cannot be due solely to the consumption of meat because plants and meat are indistinguishable in their ^{44}Ca: ^{42}Ca ratios. Therefore, we hypothesized that the ingestion of bone marrow, with as little as 1 percent trabecular (spongy) bone, was responsible for the specimen's calcium isotopic signature. This methodology will almost certainly be applied to Cro-Magnon skeletons in the future.

INTEGRATION OF CALCULUS AND MICROSCOPIC PLANT REMAINS INTO DIETARY RECONSTRUCTION

Calculus is a durable substance that has been recovered from fossils as old as the 8- to 12-million-year-old orangutan relative *Sivapithecus*. Although primarily composed of the calcified remains of oral bacteria, we know that pollen or other airborne particles inhaled through the mouth, and even tiny bits of food, can also become trapped within it. In recent years, researchers have turned to dental calculus on fossil teeth as another source of data for understanding prehistoric diet—especially its plant components. Two types of microfossils found in calculus are considered to be the most informative. The first are phytoliths, microscopic silicon-containing structures found in many plant tissues. Phytoliths in calculus need not be indicative of diet, however. They could be due to plant-based bedding, "toothpicking," or using the teeth as a vise when working wood or other materials. They could even be due to the medicinal use of plants (think of the old European practice of chewing willow bark).

The second type of microfossils, starch granules, are more informative when it comes to diet. Starch is a plant's means of storing glucose in a large, stable molecular form. Although present in a variety of plant tissues, it is particularly abundant in seeds and in underground storage organs such as potatoes. In the 1990s, I watched fitness guru Covert Bailey depict starch on a whiteboard as a string of glucose molecules, each molecule represented by the letter "G" (Bailey then referred to starch as the "G-string"). Here's the rub. All that stored glucose makes dietary starch an invaluable energy source and an important dietary component for many animals, including humans. Today it accounts for approximately 50 to 70 percent of human caloric intake!

Hardy et al. note that any starchy item eaten contains billions of starch granules; of these, only a tiny proportion become stuck in calculus, and among those that do get stuck, there is considerable morphological homogeneity, making it difficult to pinpoint the species of plant from which the starch granule has come.[27] That said, wheat, barley, and rye seeds, both wild and domesticated, tend to show a distinctive bimodal size distribution and are recognizable when large and small granules are found together. Keep in mind that even a small amount of cooking alters the chemical structure of starch, making it much more easily broken down by salivary amylase enzymes; for this reason, any recognizable starch granules recovered from calculus are probably from an uncooked source.

Thus far no identifiable starch granules from grains have been recovered from the calculus of Pleistocene Cro-Magnon specimens, although one starch granule from Abri Pataud is morphologically close to those of the water lily family. In the Holocene, however, Cristiani et al. showed that the calculus of 6,600-year-old Mesolithic hunter-gatherers from the Balkans holds starch granules of domesticated grains, indicative of trade between Neolithic farmers and the Mesolithic hunter-gatherers in the region at that time.[28]

Perhaps the most exciting revelation in recent years regarding the consumption of plants by the Cro-Magnons has been the recovery of microscopic plant remains still adhered to the working ends of several 28,000- to 33,000-year-old stone pestles from sites in Czechia, Italy, and Russia.[29] Of particular interest is one from the Grotta Paglicci in southern Italy, from a layer dated to 32,600 cal BP. Under a microscope its irregular starch granules resemble those found in acorns (*Quercus*), and others resemble granules from the root organs of cattails (*Typha*). Revedin and colleagues hypothesize that these energetically rich materials were ground into flour—a highly nutritious and easily stored commodity typically associated with agriculture. As a test of this hypothesis, they analyzed modern acorn, cattail rhizome, and wheat flour. All three of these flours are rich in carbohydrates (>50%). Acorn flour has a lower protein content than that milled from either cattail or wheat (5.0% vs. 9.1% and 11.9%, respectively), but it is fattier (8–10% vs. 2–3%). These researchers suspect that wild plant flours were a source of carbohydrates and fiber for the Cro-Magnons, and that acorn flour could have served as a complementary lipid source as well, having a fat content closer to that of olives.

It appears that the Cro-Magnons ate a diet that was diverse, healthy, and nutritionally rich, enabling them to grow to roughly the same size as people in Europe today (chapter 9). In the next chapter, I shift gears a bit, focusing on the art made by the Cro-Magnon people. This is not a complete shift in focus, however; in the early years of studying Paleolithic art, it was thought to be centered on the food quest as a form of "hunting magic."

11

Cro-Magnon Art

The time had come. Garman, Keeper of the Stories, had grown old. His joints ached and his eyes were clouded by cataracts, but his mind remained sharp. He knew all the old stories by heart and had mastered the art of telling them. Even those who had heard these stories dozens of times would sit in awed silence as he retold tales of gods and goddesses, of heroes and tricksters and villains, and tales of fabled hunts from long ago—both the glorious and the disastrous ones. Some stories had a moral, and others elicited laughter. There were stories of wise talking animals and the fate of those who foolishly rejected their counsel. He loved telling these stories, but Garman knew he would not live forever. Someone new must take over.

Six years ago Garman had chosen Mevan to be his successor. Everyone in the group knew these stories, but there was something in the way Mevan recounted the events of her day—something in how she was able to hold the audience's attention even while relating seemingly mundane events—that caught Garman's attention. Mevan would be his protégé. For six long years he coached her and listened patiently as she retold the tales. He corrected her if she left something out but did not correct her when she added a nuance that made the story better. Keepers throughout the ages had permission to add their own touches to the stories as long as the themes remained unchanged. Although Garman could no longer see Mevan's facial expressions as she told the tales, he could hear in her voice that which he could not see, and it pleased him. She would be a worthy Keeper—one who might surpass him.

On a crisp spring evening, he announced to the group that he had told his last story. Mevan was now the Keeper and would lead them deep into the cave tonight to tell her first story. Lamps and torches were lit, and everyone quietly made their way deep into the cave. Garman's youngest son helped his aged father negotiate the tight passages. When the group arrived in the chamber, torches ablaze, the paintings on the walls seemed to dance in the light, as if overjoyed to see the people once again. A hushed reverence fell upon the crowd, everyone waiting for tonight's story to begin. Mevan lifted her torch so that it shone on the painting she had chosen and began:

> Long ago, at the beginning of time, in this same valley, lived a magical shaman known as the Birdman. He was the father of our people, and we honor him tonight by telling his story.

I discussed the capacity to produce art in light of the "modern human behavior" debate in chapter 7, and in this chapter I focus on art produced by the Cro-Magnons—both in its smaller portable and in its parietal (wall) forms. The study of Paleolithic art has a long history that dates back to our old friend Abbé Breuil. Perhaps the most pervasive questions surrounding Paleolithic art concern its meaning. I'm not certain that one can pinpoint the meaning of art in our own society, so I hold little hope of understanding its function across tens of thousands of years in societies so far removed from us in time. That caveat aside, speculating about art is fun, and I'll engage in a bit of speculation here as well.

CRO-MAGNON MOBILIARY ART AND THE "FIRST FAD"

With a few key exceptions, hunter-gatherer societies are not sedentary. Instead they tend to move from one location to another, sometimes multiple times over the course of a year. We now know that places with abundant resources, such as Dolní Věstonice, were occupied year-round in the Upper Paleolithic (chapter 10), but a more mobile way of life was the rule rather than the exception for most Cro-Magnon people. Much of the art they produced was small enough to be carried from place to place without being a burden to transport. This is called mobiliary (or portable) art, and the earliest mobiliary art found in Europe comes from the Swabian Jura. This region of southwestern Germany, sometimes called the Swabian Alps, is an area of rolling tree-covered hills and gently sloped mountains. Its well-watered valleys also serve as sources for the Danube. Like the Vézère

Valley, the Swabian Jura is an ancient limestone plateau that contains caves and rock shelters ideal for Paleolithic habitation and that also provide excellent preservation of organic artifacts. The plateau is rich in Paleolithic sites, and archaeological research here in the past two decades has increased our understanding of the Cro-Magnons, especially with regard to their art.

We know that Aurignacian hunter-gatherers pierced shells and animal teeth to use as jewelry and used colorants such as ochre and manganese dioxide for decoration (chapter 8). Yet we see other art forms for the first time in the archaeological record of the Aurignacian people—including musical instruments. In 2008, the three oldest flutes yet found, all older than 35,000 cal BP, were discovered in Aurignacian contexts in the Swabian Jura.[1] The largest of the three was recovered from basal Aurignacian levels at Hohle Fels Cave in the Ach Valley on September 17, 2008 (figure 11.1). It was carved from the radius of a griffon vulture (*Gyps fulvus*).

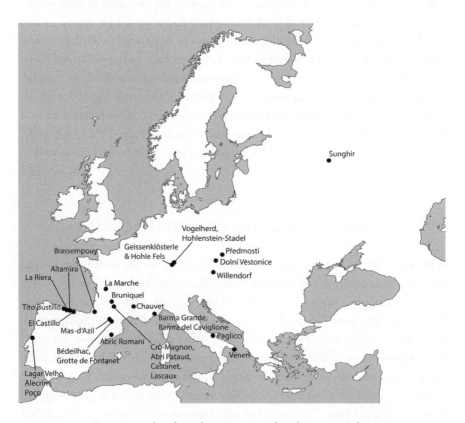

FIGURE 11.1 European archaeological sites mentioned in chapters 11 and 12.

Source: Map by author.

With a wingspan of 230 to 265 centimeters (7.5 to 8.7 feet), the hollow wing bones of these birds are ideal for making long flutes. Five finger holes were carved into the flute, and at its proximal end two V-shaped notches were cut into which the musician would presumably blow. Fine lines are also engraved into it that appear to be measurements to mark where the finger holes should be positioned.

That same summer, at the neighboring site of Geissenklösterle, a smaller three-holed flute made from a swan's radius was also recovered from Aurignacian levels. If one blows obliquely into the flute's proximal end, four notes can be made, with additional overtones possible by blowing more sharply into it. The last find of 2008, also from basal Aurignacian levels at Hohle Fels, were two small fragments of what were probably two different ivory flutes. Conard and colleagues point out that the manufacture of a flute from ivory is much more labor-intensive than making one from the wing bone of a bird because unlike long bones tusks are not hollow. One has to hollow out ivory to make a flute from it! All these instruments, and one from the nearby site of Vogelherd, point to a well-established tradition of making music earlier than 35,000 cal BP.

Figurines are also commonly discovered in Aurignacian levels in the Swabian Jura. A 5.8 centimeter (2.25 inch) ivory "Venus" figurine, dating to at least 35,000 cal BP and perhaps as early as 40,000 cal BP, was recovered from Hohle Fels Cave in 2008. She is the earliest unequivocal depiction of a human being yet discovered. Lacking a head, on her shoulders rests an off-center pierced ring (the polish on which suggests she had been strung as a pendant). Similar to later Gravettian figurines, she has big breasts, an enlarged abdomen, and a prominent vulva. Her buttocks are also pronounced; their cleft continues to the front of her pubic region. Her legs are short and asymmetrical, with the left smaller than the right. In contrast, her arms are long and curved, with carefully carved hands resting on her abdomen just beneath her breasts. Venus figurines really take off later in the Gravettian, and a common interpretation of these figurines is that they are fertility symbols.

Approximately 36 kilometers (22.5 miles) ENE of Hohle Fels, near the town of Stetten in the Lone Valley, lies the Swabian Jura site of Vogelherd. In 1931, excavations there led by Gustav Rieks recovered from Aurignacian levels multiple small, 8 to 13 centimeter (3 to 5 inch) long, lifelike carvings, most in ivory, of animals. Some are broken or unidentifiable as to species, but there are at least two mammoths, one horse, and at least one large felid, most likely a lion. Interpretations of these objects as totemic symbols are unprovable but certainly possible. That said, they could just as easily have been toys for children.

Just 1.9 kilometers (1.1 miles) upstream on the Lone River from Vogelherd lies another Aurignacian site known as Hohlenstein-Stadel. It is from this cave that one of the most famous works of Paleolithic mobiliary art was uncovered. It is known in German as the *Löwenmensch* (Lion-Man), and in good archaeological Murphy's Law form, it was discovered on the last day of excavation in September 1939. Excavations were subsequently halted due to the beginning of World War II in Europe. Lion-Man has been slowly and carefully reconstructed in the years since then. Skillfully carved from mammoth ivory, this sculpture of an upright anthropomorphic figure with the head of a lion stands 31 centimeters (just over 1 foot) tall and dates to about 40,000 cal BP (figure 11.2). His humanlike arms hang, with elbows ever so slightly flexed, at his sides; his left arm bears linear marks suggestive of tattooing or scarification. His short legs join behind what appears to be male genitalia lying beneath a triangular loincloth; the fact that he lacks a mane is not surprising (see below). The sculpture itself is realistic and beautifully detailed—even his navel is visible. Lion-Man is the oldest representation of a mythical creature found so far, and I cannot help but wonder who he was. Was he a god or a magical being? Was he a powerful shaman who had passed on to another world or a hero of legend? How widely was Lion-Man known? Would Aurignacian people in the Vézère Valley, or even as far away as Portugal, have immediately known who he was, or was his fame much more

FIGURE 11.2   The 40,000-year-old *Löwenmensch* figurine from Hohlenstein-Stadel, Germany.

Source: © Museum Ulm, Photo by Oleg Kuchar.

localized within the Upper Danube Valley? Finally, could he simply represent a person wearing a lion mask? I think that his long, catlike torso and short legs argue against this latter interpretation.

Beginning around 31,000 years ago and broadly associated with the Gravettian industry, small, stylized Venus figurines appear from France in the west all the way to Siberia near Lake Baikal in the east, and they remained popular for close to ten thousand years (I refer to them as the first fad). The figurines depict obese or pregnant females (hence the Venus moniker). Their breasts and buttocks tend to be large, but their facial features are minimal, their arms are usually tiny or even absent, and they often have stylized legs. Some are nude, with prominent vulvae, and others are wearing at least some clothing. The Venuses are made from a variety of different materials. Two well-known examples are shown in figure 11.3. The Dolní Věstonice Venus, discovered in 1925, is made from fired clay mixed with ash—giving her a rich black color. She is the oldest fired clay art object known. Small slits for eyes are her sole facial feature, and she lacks arms.

FIGURE 11.3 Two 31,000-year-old (Gravettian industry) Venus figurines. On the left is the Dolní Věstonice Venus. On the right is the Venus of Willendorf.

Source: (*Left*) From the collections of, and courtesy of, the Moravian Museum/Historical Museum/Anthropos Institute. (*Right*) Courtesy of Erich Lessing/Art Resource, New York.

The Venus of Willendorf, discovered in Austria in 1908, is carved from soft lime-stone. She is either wearing a headpiece, perhaps made of shells, or her hair is elab-orately styled. None of her facial features is indicated, and her tiny, waif-like arms lie across her bulbous breasts. Other Venus figurines are carved from materials as diverse as black jet gemstone, mammoth bone, and ivory. Given the emphasis on the figurines' secondary sexual characteristics, they have long been interpreted as fertility symbols. Other interpretations range from representing an Earth god-dess, to women creating three-dimensional self-portraits of their own bodies, to "paleo porn." Whatever their function(s), the fact that this cultural tradition continues over such a broad swath of Eurasia for so many thousands of years is amazing to me. I can think of few cultural motifs today with such longevity, with the possible exception of the use of arrows in signage to indicate direction.

Some mobiliary pieces from this period look more like portraiture (figure 11.4). The so-called Brassempouy Venus, discovered in 1892 in southwestern France, is a realistically portrayed female face carved from mammoth ivory. She appears to

FIGURE 11.4 Two Upper Paleolithic facial portraits. On the left is a front view of the Venus of Brassempouy. On the right is the Dolní Věstonice XV portrait head.

Source: (*Left*) © RMN-Grand Palais/Art Resource, New York. (*Right*) From the collections of, and cour-tesy of, the Moravian Museum/Historical Museum/Anthropos Institute.

be wearing a headdress. Her face has delicate features, including a slight nose and gentle brow, but no mouth is indicated. The Dolní Věstonice XV portrait head, discovered in 1936, is carved from mammoth ivory, and is about 5 centimeters (2 inches) in height. There is asymmetry evident between the left and right sides of the carved face. In 1949, the Dolní Věstonice 3 burial was discovered nearby. This skeleton, found beneath two mammoth scapulae, is that of a woman in her late thirties or early forties who suffered paralysis of the left side of her face, presumably due to a childhood injury. This paralysis made her face asymmetrical, leading some researchers (most notably Bohuslav Klíma) to speculate that the Dolní Věstonice XV portrait head is a realistic depiction of the Dolní Věstonice 3 woman! However, a recent computer-aided facial reproduction of DV 3 takes into account her facial asymmetry, and these researchers argue the degree of face asymmetry in life that would have been obvious to others is subjective.[2]

Some of my favorite pieces of Upper Paleolithic mobiliary art are artistic works on utilitarian objects. Many atlatls recovered from Paleolithic contexts are beautifully decorated (chapter 10). One of the most famous is a 15,000-year-old example from Mas-d'Azil Cave in the French Pyrenees. Made from a single reindeer antler and with 32 centimeters (a little over 1 foot) of its length preserved, it features what looks to be a wild goat giving birth (figure 11.5). A bird (perhaps a crow) sits atop of whatever is emerging from the animal's rear end (corbins are known to eat the caul associated with mammalian births). The bird's tail serves

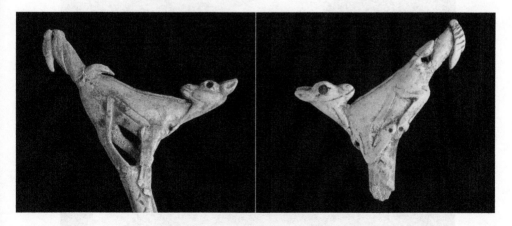

FIGURE 11.5 Views of two atlatls featuring what appears to be a wild goat giving birth. The one on the left is from Mas d'Azil; the one on the right is from Bédeilhac.

Source: © RMN-Grand Palais/Art Resource, New York.

as the atlatl spur. As with so many Paleolithic works of art, I cannot help but wonder if this piece depicts a well-known story or legend. The existence of a second, less complete atlatl featuring the exact same motif—this one from the site of Bédeilhac, some 28 kilometers (17.5 miles) to the southeast of Mas-d'Azil—lends credence to the notion that the bird-on-goat giving birth theme, whatever its meaning, was well-known in the region; although it is also possible both pieces were made by the same artist.

Another of my favorite mobiliary art pieces is the "leaping horse" from Bruniquel in southern France, a piece carved from reindeer antler depicting a horse in mid-leap (figure 11.6). Its front legs are tucked under its chest, and its hind limbs are at full extension. Its nostrils, eyes, ears, mane, tail, and even the hairs on its body are all carved in exquisite detail. Long thought to be an atlatl because its grip end is perforated as if for a strap, its use as an atlatl would be problematic because it lacks a spur. I suspect it had been intended to be used as an atlatl but that its spur was broken either during its manufacture or use.

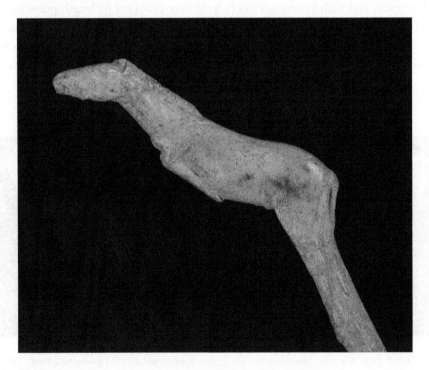

FIGURE 11.6 The "Leaping Horse" atlatl from Bruniquel.

Source: Courtesy of the Musée d'Archéologie Nationale/Art Resource, New York.

PARIETAL ART

In the public imagination, no art form is more associated with the Paleolithic than the famous cave paintings of Europe—art that in many cases was so skillfully done it was originally thought to have been a forgery. I discussed examples of European cave art that were probably the work of Neandertals in chapter 7. Here I focus on the Cro-Magnons' cave art. The oldest examples of parietal (wall) art in caves that are almost certainly the work of Cro-Magnons date to the Aurignacian, which began about 43,000 years ago. This art is engraved or painted with organic and inorganic pigments—red and yellow pigments are almost always inorganic, and most, but not all, black pigments are charcoal. Recall that animals are frequently represented in Aurignacian mobiliary art. With the possible exception of Chauvet Cave, relatively few animals are portrayed in Aurignacian-age parietal art, and those that are portrayed are done only in simple outline form. More common Aurignacian parietal art motifs include stencils of hands and handprints, painted dots and other geometric symbols, and carved vulvar forms.

In 2012 the British archaeologist Alistair Pike and his colleagues reported Uranium-series dates on calcite deposits either partially covering, or underneath, parietal art at eleven cave art sites in northern Spain, including the UNESCO World Heritage sites of Altamira, El Castillo, and Tito Bustillo.[3] They found multiple examples of cave art dating to the Aurignacian. The two oldest were on the *Panel de las Manos* (Panel of Hands) at El Castillo. The first is a dated calcite deposit overlying a large red stippled "disc" design; its age is 41.4 ± 0.57 ka (note that this early, the disc's age could correspond to the Châtelperronian and thus *could* be the work of Neandertals). The second dated calcite layer overlies a red negative hand stencil, probably made by blowing powdery red pigment over a hand placed on the wall, and underlies a yellow bison outline. It dates to 37.63 ± 0.34 ka.

Other examples of Aurignacian-age cave art in Spain dated by Pike and colleagues include a red claviform (club- or P-shaped) image from Altamira (36.16 ± 0.61 ka), a large red disc from the *Galería de los Discos* (Gallery of Discs) at El Castillo (34.25 ± 0.17 ka), and an anthropomorphic (almost "stick like person") figure from the *Galería de los Antropomorfos* (Gallery of Anthropomorphs) at Tito Bustillo (30.8 ± 5.6 ka).

Examples of Aurignacian parietal art have been found in France as well. In 2012, Romain Mensan and colleagues reported the discovery of art on a giant (one ton) bloc that had fallen from the roof at the site of Abri Castanet (located between Les Eyzies and Montignac) over 30,000 years ago.[4] The bloc was removed in 2007, during the first archaeological work to take place at the site since Didon's excavations in 1912. The underside of the bloc, which had been the shelter ceiling, was in direct contact with an Aurignacian level that was radiocarbon dated using ultrafiltration on bone to about 32,400 BP (uncalibrated). Its ceiling side was engraved and stained with red ochre. The most striking engraving is that of a circle, about 13 centimeters (5 inches) in diameter, into which enters a slightly curved line some 9 centimeters (3.5 inches) in length and 1.5 centimeters (0.5 inch) in width. Mensan and colleagues are uncertain about how to interpret this symbol, pointing out that many ovoid engravings from Aurignacian contexts are interpreted as vulvae; this one, however, is much more circular and includes the enigmatic curved line.

The most striking examples of potentially Aurignacian-age parietal art are the fantastic cave paintings at Chauvet Cave in the Ardèche Valley of southern France, discovered in December 1994. It is sometimes called Chauvet-Pont d'Arc Cave because it lies close to a natural arch bridge across the Ardèche. In addition to both positive and negative handprints and the red dot motifs one sees in other Aurignacian-age art, nearly three hundred depictions of animals are rendered on the cave's walls. Some of these depictions are engravings, some simple line drawings, but most of the painted images are so realistic and detailed that they are recognizable as to species. The most frequently rendered animal at Chauvet is one that is rare at other cave art sites: the woolly rhinoceros, which accounts for nearly one-quarter of all of the depicted animals. At times they are drawn with exaggeratedly long front horns. They and many other animals are shown in overlapping form, suggestive of either great numbers or of movement. Cave lions are the next most frequently depicted animal, making up approximately 17 percent of the animal depictions (figure 11.7). Both male and female lions are represented in the paintings, with the external genitalia visible in some renderings. In these paintings, males do not have manes. Unlike lions today, male cave lions lacked manes, a fact we know only because of our ancient cousins' silent witness.

The Chauvet team maintain that the art dates to two periods, the first from 37,000 to 33,500 years ago (placing it solidly within the Aurignacian), and the second from 31,000 to 28,000 years ago, making it Aurignacian or Gravettian.[5] These ages remain controversial because the vast majority of radiocarbon dates are on hearths that may have nothing to do with the paintings on the walls above them, and the direct dates of the paintings themselves are on charcoal, a notoriously difficult substance to date.[6] In addition, only one laboratory dated the charcoal, so none of the dates is corroborated. Finally, much of the art in Chauvet Cave is so stylistically similar to later Gravettian-aged, and even post-Gravettian-aged, art elsewhere in France and Spain that many archaeologists argue it is unlikely that the magnificent Chauvet paintings are of Aurignacian age.

FIGURE 11.7   Panel of Lions from Chauvet Cave.

Source: Courtesy of HIP/Art Resource, New York.

The art from the Aurignacian is both fascinating and important, but the heyday of Upper Paleolithic parietal art was much later in time, in the Magdalenian, which began about 18,000 years ago. Most of the paintings from the world-famous cave art sites of Altamira in Spain and Lascaux in France are dated to the Magdalenian. The artists in these caves created the illusion of three-dimensional animals through several techniques: (1) using the contours of the cave to "suggest" the shapes of animals, (2) applying the technique of "vanishing point" perspective, and (3) employing a high degree of shading variation, often scraping away at the wall before painting to create a whiter surface for greater contrast.[7] The result of these techniques is evident in the Lascaux painting sometimes referred to as the "cheval chinois" (Chinese horse), so named because of its uncanny resemblance to the wild Przewalski's horse found today on the Central Asian steppes (figure 11.8).

In Paleolithic cave art, especially at Lascaux, but to some extent at sites like Altamira and Chauvet, many animals are depicted in what Breuil called a "twisted perspective" in which the animal is shown in lateral (side) view, but the antlers or horns are depicted as if the head is facing the viewer. In our own age, Charles M. Schulz employed this method to represent the bill on Charlie

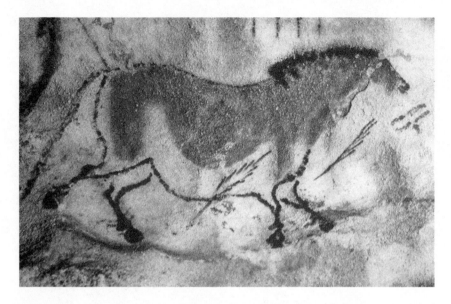

FIGURE II.8 Painting of a "Chinese horse," Lascaux Cave.

Source: © Alinari Archives/Art Resource, New York.

Brown's baseball cap. When Charlie is facing the viewer, the bill of his cap is drawn out to the side so that it remains visible. As a young child, I misinterpreted this visual cue and would put my baseball cap on sideways when I pretended to be Charlie Brown![8] An additional aspect of the twisted perspective is that the hooves of some animals drawn in profile are depicted in such a way that their hoofprints show. Many scholars have hypothesized that the purpose of this perspective is to provide the viewer with the most relevant information about the animal in question. If the animal was depicted purely from the front, one could not discern the shape of its body. When purely in profile, one could not see both horns/antlers, nor could one make out the all-important hoofprints of the animals in question. (Similarly, it allows one to know that Charlie Brown is wearing a baseball cap and not a beanie!)

With the exception of La Marche Cave in western France, which features engravings of human faces, humans are rarely depicted in the parietal art of this period. Another exception happens to be one of my favorite pieces of parietal art—the so-called Birdman from Lascaux (figure 11.9). In this scene, a bison has been disemboweled by an atlatl dart still stuck in its side. It appears to have responded to this injury by goring a humanlike stick figure who has a bird head and fingers resembling bird feet, hence the "Birdman" moniker. As a result,

FIGURE 11.9   Enigmatic Birdman painting, Lascaux Cave.

Source: © Ministère de la Culture/Médiathèque du Patrimoine, Dist. RMN-Grand Palais/Art Resource, New York.

Birdman is falling backward; perhaps his most notable feature is his erect penis. At his feet lies what appears to be a second atlatl dart. Nearby a bird, whose face is similar to that of the Birdman, sits atop a stick. Some have interpreted this bird-stick as the hunter's atlatl; as previously discussed, many atlatls are decorated with animal motifs. To the left just beyond this surreal scene a rhinoceros appears to be moving away from the attack, although it is not certain that the rhinoceros is contemporary or is associated with the birdman and bison. Who was this so-called Birdman? Was he a shaman or a mythical creature? Does this image tell a "true" story—a hero or creation myth—or does it depict a shamanistic vision? Like the goat giving birth motif, I wonder if this scene was a well-known story at the time that anyone from the surrounding area would immediately recognize.

This brings us back to the thorny issue of the meaning of cave art. I do not believe that art created across Europe for tens of thousands of years had a singular function or meaning across all that space and time, and the majority of my archaeological colleagues would agree with me. The early days of Paleolithic art study, however, were dominated by Abbé Breuil, who interpreted the art as "hunting magic." The problem with Breuil's hypothesis is that the animals most often depicted in cave art were not those most frequently eaten. As an obvious example, I assure you that the people who lived in and around Chauvet Cave did not base their subsistence on the woolly rhinoceroses and cave lions so prevalent in their art. Focusing on these dangerous animals is one surefire way to leave no descendants in the next generation or to win the Darwin Award, as people so frequently joke these days. Similarly, we know from zooarchaeology that the people of Altamira subsisted primarily on red deer—not the bison so often painted on the walls and ceilings of the cave. Likewise, when it came to meat, the people of Lascaux ate almost nothing but reindeer, yet horses are the most frequently painted animal, comprising 60 percent of the total images—and not a single reindeer is depicted on the cave's walls! If these paintings had truly been done to ensure the success of the hunt, wouldn't they more often than not depict the animals most frequently hunted?

Many writers have focused on the ritual aspect of cave art. The fact that so much cave art is found in hard-to-access areas deep within the dark zone—areas where artificial light would be needed both for the creation and viewing of the art—is certainly consistent with a ritual purpose. We know the Cro-Magnons had multiple ways of lighting the dark—animal fat lamps made from stone were recovered from deep within Lascaux Cave, and black marks where torches were revived by rubbing them on the cave walls are found throughout all caves containing decorated walls. I would certainly not rule out a ritual function for the art found in such hard-to-reach places, but part of the reason this art survives is because of its inaccessibility. We've learned from the experience of Altamira and Lascaux that people breathing around cave art leads to its eventual destruction. Perhaps even more important, in 2021 the French geoarchaeologist Laurent Bruxelles showed that caves housing bat colonies are devoid of cave art. The heat from the bats, the carbon dioxide they exhale, and their excrement together create a microenvironment within the cave that quickly erodes away the cave floors and walls, leading to the rapid destruction of the art. In other words, these deep recesses of caves, many of which were cut off from the remainder of the cave system due to rock falls (or underground aquifer rise), are among the few places where well-preserved cave art is expected to survive! There was almost certainly cave art throughout the caves inhabited by the Cro-Magnons, including in more accessible areas—art that has long since vanished due to the presence of humans or bats!

Of course, other functional explanations for parietal art are not mutually exclusive. Recall that in many places the images of the animals overlap or are drawn as if multiple individuals are lined up in rows. Some animals are even depicted with additional legs. These effects help create a feeling of motion, and when the images are viewed by the flickering light of a torch or lamp, the illusion is even further enhanced. In *Growing Up in the Ice Age*, the archaeologist April Nowell seized upon this phenomenon to argue that Paleolithic cave art played a central role in that most human of activities—storytelling.[9] Her idea does not preclude a ritual function for these stories, but one can also imagine, as she does, that at least some stories were meant to teach children important lessons. I cannot help but wonder if they often ended as German fairy tales do: "And if they hadn't died, they'd still be alive today." In other cases, the stories could have been comedies rather than tales with a religious, moral, or tragic theme, and Nowell points out that in some cave art the animals on the walls display wry smiles and emotive eyebrows!

I remain agnostic as to the function(s) of Paleolithic cave art, but I also wonder if we lose something when we try to analyze it too carefully. To paraphrase E. B. White and K. S. White: Paleolithic art can be dissected, as a frog can, but it dies in the process.[10] The art is breathtakingly beautiful and, although it may be sacrilege for me as a biological anthropologist to say this, it gives us greater insight into the lives of the Cro-Magnons than the mere study of their skeletons ever can.

In the final chapter, I examine how the Cro-Magnons dealt with climate change—a natural phenomenon that probably killed many thousands of them, and one that is affecting our own lives today.

12
Cold Comfort

Climate change and its adverse effects have been upon us for decades now, with record summer heat in both the northern and southern hemispheres, unpredictable winter weather, and an increased frequency of tropical cyclones. For example, the summer of 2003 brought record high temperatures to Europe, along with an estimated 30,000 heat-related deaths on the continent, roughly half of which were in France. A temperature anomaly map for that summer shows that France was much, much hotter than normal, but other parts of the continent were also sweltering—I know because I was in Portugal doing archaeological fieldwork. The 2003 temperatures were not as unseasonably hot in Portugal as they were in France; after all, Portuguese summers are typically hotter than French ones. Nonetheless, the heat that year was oppressive, and it is not a summer I will soon forget.

In 2003 I was on a team excavating a rock shelter in central Portugal called *Abrigo do Alecrim* (Rosemary Shelter), named for the rosemary bush growing out of a crevice in its boulder roof. The shelter had been discovered and initially explored at the end of the previous summer's field season by two of my former students, Vance Hutchinson and Linda Tuero Lindsley. For the 2003 season, the archaeologist Francisco "Chico" Almeida was brought in to lead the excavation.

Abrigo do Alecrim lies in the canyon portion of the Lapedo Valley of central Portugal, just 300 meters (0.2 miles) NNE of the Lagar Velho rock shelter that yielded the skeleton of the presumed hybrid child (chapter 5). Our team would spend a few more seasons at Alecrim, with Telmo Pereira taking charge of the site upon Almeida's emigration to Australia (figure 12.1). Alecrim is a tiny shelter, and we only excavated a 2 × 2 meter (6.5 × 6.5 foot) pit, reaching bedrock at the end of our final season in 2007.

FIGURE 12.1 The 2005 field season at Abrigo do Alecrim. From left to right: the author, Telmo Pereira, and Francisco Almeida.

Source: Photo by Vance Hutchinson; collection of author.

Excavations at Alecrim began in earnest in the summer of 2003, and it was the hottest year we experienced. On August 1, 2003, the temperature spiked at 41.4° C (106.5° F) in Leiria, the city closest to our site. I grew up in Louisiana, and I'm telling you it was *hot* in Portugal that year. Forest fires raged across the country, and late one morning we began to smell wood smoke while we were excavating. Suddenly a searing hot wind was whipping through the canyon, bringing millions of floating paper-thin pieces of ash with it. A forest fire was on the move toward us. We decided to quickly cart out as much of our equipment as we could, leaving the rest as deep in the shelter (and therefore as far from potential fuel) as possible. Having abandoned the valley for the day, we spent the evening safely ensconced miles away, but we worried all night that fire had engulfed our site.

The next morning brought much less smoke, but we were unsure of what we would find at Alecrim. Tentatively we drove into the Lapedo Valley from the burned east side via the village of Caranguejeira. Much of the vegetation, a Mediterranean mix of pine and scrub oaks, on both sides of road was charred, but about 500 meters

(0.3 miles) from Alecrim we no longer encountered evidence of burning. At some point during the previous day, the wind had shifted, sparing our site from harm.

Alecrim faces east, and the sunlight would first reach us around 8 a.m. and stay with us for much of the day as we excavated—the shelter had no natural shade until about 6 p.m. We tied a tarp nearly vertically between two nearby trees in an attempt to block out some of that intensely hot sun, but in 2003 there was no escaping the heat. From a temperature perspective, it was an absolutely miserable season. On the bright side, our little site yielded thousands of pieces of faunal bone, much of it burned in a simple hearth, as well as hundreds of lithic artifacts, most made on local quartz or quartzite. Like the nearby Lagar Velho, most of the lithics from Alecrim are not diagnostically Upper Paleolithic, with over 80 percent manufactured on flakes. Given the shear density of artifacts, Alecrim seems to have been intensely occupied, although probably by a small group (a maximum of ten to twelve people) considering its size. They may have used the site more than once over several hundred years; our AMS radiocarbon dates on faunal bone ranged from 20,500 ± 150 to 21,800 ± 170 (uncalibrated) years BP. These dates suggest the main hearth dates to the Proto-Solutrean, corresponding to the Last Glacial Maximum (LGM), the coldest period of the past 120,000 years.

Given the climate when Alecrim was occupied, the sun that cursed us during the unseasonably hot summer of 2003 was almost certainly a godsend to its chilled-to-the-bone Proto-Solutrean occupants, who probably chose to stay there *because* of the amount of sunlight it received. Living in caves and shelters that receive a lot of sunlight is just one of many tricks the Cro-Magnons of the Last Glacial Maximum employed to keep warm. As Aesop said, necessity is the mother of invention, and a number of technological innovations for keeping warm make their first appearance, or become widespread, during the LGM. Many of these innovations are still used today. Although largely abandoning northern Europe during the LGM, the Cro-Magnons nonetheless managed to survive in southern refugia such as Iberia, and they expanded northward again as temperatures rose about 17,000 years ago, although the area was still under glacial conditions. They survived through the end of the Ice Age and experienced the global warming that marks the beginning of the Holocene (the current epoch) about 11,700 years ago. It may seem counterintuitive, but in myriad ways the warming and reforestation of Europe with the onset of the Holocene was a more difficult challenge for them than the extreme cold of the LGM.

In this chapter I explore the Cro-Magnon response to climate change, which offers lessons all too relevant for us today as we negotiate life on a rapidly warming planet. But first I provide a primer on the study of ancient climate (paleoclimatology).

EARTH'S CHANGING CLIMATE

Earth is a dynamic planet in every sense of the word, and its climate is no exception. At various points in history, Earth's climate has been affected by extraneous factors. For example, 66 million years ago an asteroid or comet struck what is now the Yucatán, setting off a global winter and bringing about the extinction of nonavian dinosaurs and countless other species. However, Earth is also tectonically active, and this affects our climate as well. It is hypothesized that thermal vents deep in the Atlantic released huge amounts of methane some 10 million years after the dinosaur extinction, 56 million years ago, setting off the Paleocene-Eocene Thermal Maximum (PETM), a 170,000-year-long period of global warming.

Earth's continental plates slowly grind past each other, move away from one another or ride on top of or become subducted beneath each other—dynamics that give rise to earthquakes, mountain uplift, and volcanism. These tectonic changes can also alter the climate. About 335 million years ago all of the continental landmasses we know today were joined together in a giant supercontinent known as Pangaea. It may not surprise you to learn that at the equatorial center of this vast continent lay a hot, arid, nearly lifeless desert. All of the rain fell long before any clouds reached deep into the supercontinent's hot interior, and the Earth was so warm that there were no polar ice caps! Much later, around 92 million years ago, when dinosaurs roamed the planet during what is called the Cretaceous Hot Greenhouse Earth, forests thrived across Antarctica, and there were no polar ice caps then either. This particularly long warm phase may have been caused by widespread volcanic activity releasing massive amounts of CO_2 into the atmosphere.

Early in the Age of Mammals, about 49 million years ago during the Eocene epoch, the Drake Passage between South America and Antarctica opened, allowing cold ocean currents to circulate completely around Antarctica. This began a long general cooling trend that lasted through the end of the Pleistocene. At nearly the same time as the opening of the Drake Passage, the Indian subcontinent began to collide into Asia proper, ultimately leading to the uplift of the Himalayas and the Tibetan Plateau. By 10 million years ago, the Himalayas had become the highest mountains in the world. Their presence altered the climate of Central Asia, making it much drier and colder. In turn, this may have played a role in intensifying the Pleistocene glaciations—the last of which is the one the Cro-Magnons lived through.

I have focused almost exclusively on events in the Pleistocene epoch in this book. The Pleistocene itself began about 2.58 million years ago and is marked by long cold periods known as *glacials*. Glacials are punctuated by shorter, warmer periods known as *interglacials*. In a very real sense, what we call the Holocene (the last 11,700 years) is just the most recent Pleistocene interglacial. The climate within a glacial phase is not monolithic; shifts can occur on the scale of just a few thousand years and, frighteningly, sometimes on the scale of decades. Colder phases within glacials are known as *stadials*, and slightly warmer (but still glacial) phases are known as *interstadials*. For most of the Pleistocene, glacial-interglacial cycles were about 41,000 years long, but between 1.25 million and 700,000 years ago the cycles lengthen to around 100,000 years.

What drove these glacial-interglacial cycles? The continents are more or less in their current positions in the Pleistocene, so tectonic movement is not a factor (although the fact that so much of the Earth's land area is in the northern hemisphere plays a role). We are also fortunate that no large celestial bodies hit the Earth during this time. What about carbon dioxide (CO_2)? Everyone knows that we are currently witnessing rapidly rising atmospheric CO_2 levels due to humans' unprecedented burning of fossil fuels, especially over the past 150 years. (Some scholars refer to this time of humanly induced climate change as the Anthropocene.) We are also painfully aware of the causal relationship between atmospheric CO_2 levels and the greenhouse effect on climate change. That said, CO_2 is not the only nor probably even the primary factor driving Pleistocene glacial-interglacial cycles.

The onset of ice ages versus periods of global warming in the Pleistocene were primarily driven by interactions among three so-called Milankovitch cycles, named for their discoverer, the Serbian climate researcher Milutin Milankovitch (1879–1958). The first is *eccentricity*, a 100,000- to 400,000-year shift during which the Earth's orbit becomes less circular and more elliptical. A more elliptical orbit, caused by the gravitational pull of the gas giants Jupiter and Saturn, affects the length of the seasons.

The reason we have seasons at all is due to what drives the second Milankovitch cycle—*obliquity* (or tilt). When the northern hemisphere is tilted toward the sun during the day, it is our summer. When the northern hemisphere is tilted away from the sun during the day, it is our winter and the southern hemisphere's summer. The fact that the Earth's axis is tilted at all relative to the plane of the solar system is due to a massive collision of Earth with a Mars-sized celestial body about 4 billion years ago—a cosmic event that led to the formation of our moon and left the Earth's rotational axis tilted about 23.5° from the vertical. The reason obliquity qualifies as a Milankovitch cycle is that this tilt continues to vary between 21.5° and 24.5° every 41,000 years or so—the gravitational pull of the moon prevents it from exceeding these limits. Given this cycle's length, it has been implicated in the 41,000-year-long glacial-interglacial cycling of the early Pleistocene. The shift to the 100,000-year-long cycles is more difficult to explain; here CO_2 levels may have played a role.

The final Milankovitch cycle is *precession*, a change in the orientation of Earth's axis of rotation akin to the wobble of a spinning top. Every 21,000 years or so, primarily due to the gravitational pull of the sun and the moon, Earth's rotational axis completes a full revolution, tracing an imaginary cone in space. Precession itself affects so-called Heinrich events, named for their discoverer, the German geologist Hartmut Heinrich. These events occur on average every 10,000 years or so and involve massive calving off of icebergs from ice sheets into the northern Atlantic. This results in a decrease in both the northern Atlantic's salinity and temperature, and it also alters ocean currents such that the Gulf Stream deviates much farther south, ultimately leading to a cooler Europe.

Given the asynchrony among the three Milankovitch cycles, it is a chaotic system, and glacial periods may differ in length. Although the amount of solar energy reaching Earth is more or less constant, the Milankovitch cycles affect where on Earth that energy is received, as well as its timing. For example, shifts in planetary tilt affect the amount of sun received near the poles versus near the equator, and changes in precession affect the amount of summer sun received at high latitudes. During the Pleistocene, colder glacial periods occurred when the combined action of these cycles caused the northern hemisphere to receive less solar energy in the summer, allowing the northern ice sheets to expand. In turn, these large white sheets of snow and ice reflect more of the sun's energy away from Earth, leading to further cooling of the planet and creating a positive feedback loop that expands the ice sheets farther still. During the Pleistocene, the northern hemisphere was more important with regard to global cooling because there is more land area in the northern hemisphere and ice sheets form more readily on land than over the ocean. Also, the Himalayas may have played a role by keeping warm Indian Ocean air out of central Asia.

Much of what we have learned about climate change in the Pleistocene has come from the study of deep cores drilled into either the ocean floor or into ancient ice sheets such as those covering Greenland and Antarctica. These cores provide a record of the oxygen deposited long ago. In particular, climate scientists are most interested in the changes in the ratio between two stable isotopes of oxygen, ^{18}O and ^{16}O, over time. Deviations from the "standard" ratio are denoted as $\delta^{18}O$. This long ago deposited oxygen is studied by first extracting it either from invertebrate shells in ocean floor deposits or from within the ice sheet, which is itself marked by discernable annual layers. ^{16}O is the most common oxygen isotope found in nature, and it accounts for more than 99.7 percent of all oxygen atoms. In contrast, the heavier ^{18}O isotope makes up only 0.2 percent of the Earth's oxygen atoms.

Oxygen also makes up most of the mass of water (H_2O) on Earth, and it is water's physical properties that help us reconstruct paleoclimates. Specifically, because ^{18}O is the heaviest stable isotope of oxygen, water molecules containing ^{18}O tend to evaporate less readily but condense more quickly than water molecules containing the lighter, much more common ^{16}O isotope. Imagine we are in warm equatorial waters. As the sun evaporates water from the sea, water molecules containing ^{18}O are slightly underrepresented in the water vapor, although the heavier isotope is still present. However, as that water vapor rises in the atmosphere or moves away from the equator, it cools, condenses, and ultimately falls as precipitation. Because water molecules containing the heavier ^{18}O isotope tend to condense more readily than molecules containing the lighter ^{16}O isotope, the first precipitation to fall, generally at low- to mid-latitudes, will be enriched in ^{18}O. For this reason, water in tropical oceans is slightly ^{18}O-enriched. As the water vapor gets closer and closer to the poles, however, precipitation such as snow will show depleted heavy oxygen (^{18}O) because most of the heavier isotopes in the water vapor had already earlier fallen as rain. This means that the water "locked up" in polar ice sheets has lower ^{18}O levels than the oceans.

Climate scientists have measured $\delta^{18}O$ in cores drilled into ocean floor deposits around the globe or in ice sheets and now have a documented record for the last several hundred thousand years during which Earth has gone through repeated cold periods: glacials, in which huge swaths of the northern hemisphere are covered in ice, and warmer periods (interglacials). These climatic phases are referred to as Marine Isotope Stages (abbreviated as MIS in table 12.1). Our current interglacial (the Holocene) largely corresponds to MIS 1, and broadly speaking, the last glacial period spanned MIS 2, 3, and 4. The last fully interglacial period prior to the Holocene was MIS 5e, which ended about 119,000 years ago.

During glacials cool air extends much closer to the equator, and more ^{18}O-enriched precipitation falls at low- to mid-latitudes. This means the high-latitude snow forming ice sheets at the poles has even less ^{18}O than what we see today. All that ^{18}O-depleted water being locked up in ice sheets means that the oceans during glacial times are slightly ^{18}O-enriched. Thus, by examining $\delta^{18}O$ deep beneath the seabed, we can tell whether more ice was locked up in ice sheets than today if the oxygen extracted from marine shells made of calcium carbonate $(CaCO_3)$ is rich in ^{18}O. We must correct for bias in these data, however, because shells have higher ^{18}O than the seawater in which they are found, and shells formed in colder waters tend to have an even greater proportion of ^{18}O than shells found in warmer climes (i.e., this is a nonlinear relationship).

TABLE 12.1 Timing of the Marine Isotope Stages of the
Holocene and Late Pleistocene

Epoch	Stage	Marine isotope stages	Significant events	Dates
Holocene	Meghalayan	MIS 1 (Begins ca. 14.2 ka)		4.2 ka
	Northgrippian		8.2 ka event	8.276 ka
	Greenlandian		9.2 ka event	11.7 ka
Pleistocene	Late Pleistocene (Stage 4)	MIS 1	Younger Dryas	
		MIS 2 (Begins ca. 34.6 ka)	Last Glacial Maximum	
		MIS 3a (Begins ca. 38.8 ka)		
		MIS 3b (Begins ca. 44.2 ka)		
		MIS 3c (Begins ca. 56.9 ka)		
		MIS 4 (Begins ca. 72.4 ka)	Beginning of Last Glacial Period	
		MIS 5a (Begins ca. 85.1 ka)		
		MIS 5b (Begins ca. 92 ka)	Cold/Dry Phase	
		MIS 5c (Begins ca. 106 ka)		
		MIS 5d (Begins ca. 116.5 ka)	Cold/Dry Phase	
		MIS 5e (Begins ca. 132 ka)	Last Full Interglacial	
	Middle Pleistocene (Chibanian)			129 ka

During interglacial periods, ice sheets melt, releasing huge amounts of ^{18}O-depleted fresh water into the high-latitude oceans, restoring lower preglacial ^{18}O:^{16}O ratios and decreasing ocean salinity. The study of these ratios has revolutionized climate science, and we have come to see a complex picture for Pleistocene climate. Milankovitch cycles; shorter, more regional phenomena such as Heinrich shifts; and even major swings in temperature on the order of decades known as Dansgaard-Oeschger (DO) events are all part and parcel of the types of climatic and environmental changes dealt with by our Pleistocene ancestors.

THE LAST GLACIAL MAXIMUM

The Last Glacial Maximum (LGM), about 26,500 to 19,000 years ago, falls in the second half of MIS 2 and is the coldest phase of the past 120,000 years. During this phase, a huge glacial ice sheet covered a swath of northern Europe, including Scandinavia, the northern British Isles, and the entire Baltic region (figure 12.2). The Alps were also buried under a thick layer of ice. The paleoclimatologist Jessica Tierney and her colleagues estimate that global temperatures were on average 6° C (11° F) cooler than today during the LGM.[1] To put this in perspective, the average global temperature during the twentieth century was roughly 14° C (57° F), whereas during the LGM the average temperature was about 8° C (46° F). Cooling in the LGM was not, however, evenly distributed across the globe. The high latitudes experienced colder temperatures still; in fact, western Europe had some of the steepest temperature gradients in the world during the LGM. The deviated southern Gulf Stream kept the south of Portugal temperate (akin to southern England today), whereas just 600 kilometers (370 miles) to the north in Galicia temperatures were similar to what one would today experience in northern Finland. Temperature changes in the tropics were much more muted— around 4° C (7° F) cooler than they are now—but changes in aridity were common in the tropics.

Due to these colder temperatures, the ecology of Europe was drastically different than it is today. Aside from treeless tundra in the far north and at high altitudes in mountains such as the Alps, Europe today would primarily revert to forest, with boreal (conifer) forests in the north, mixed conifer and broadleaf forests across its midsection, and Mediterranean oak forests in the south. During the LGM, however, treeless tundra could be found as far south as southern Spain, Italy, and Greece, and a vast grassy steppe (the "Mammoth

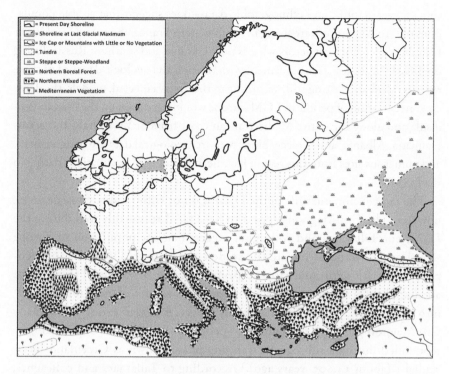

FIGURE 12.2   Map of European ecological zones during the Last Glacial Maximum.

Source: Map by the author following Carl Diercke, Wiebke Gehring, and Thomas Michael, *Diercke Weltatlas* (Braunschweig: Westermann, 2008).

Steppe") extended from the middle and lower Danube Valley east and north into what is now Russia and Ukraine (chapter 10). Likewise, boreal forests that today dominate the landscape of Finland and northern Sweden were found in the LGM as far south as the southern Bay of Biscay, parts of the interior of southern Spain, and even in what is now Bulgaria and the greater Po Valley of Italy, including Venice.

How did the Cro-Magnons cope with the colder temperatures of the LGM and its long-reaching environmental effects? It seems that they largely abandoned the northernmost parts of Europe. Archaeological data support this; few archaeological sites dating to the LGM have been found in the southernmost British Isles, the Netherlands, Belgium, northern France, and northern Germany in the west, or eastward across Poland, Belarus, and northern Russia. These areas of windswept steppe tundra lay close to the European ice sheet and would have been bitterly cold year-round. Computer modeling by Tallavaara et al.—using climatic, geographic, and ethnographic data—does suggest that sizable populations of Cro-Magnons were living in southwestern France, the middle and lower Danube Valley, and east across what is now Ukraine and southern Russia during the LGM.[2] This, too, is in agreement with the rich archaeological record of these regions. More interesting still, according to the Tallavaara et al. models, the densest concentration of people in LGM Europe would have been on the Iberian Peninsula and along the entire Mediterranean coast from Spain to Italy and across Dalmatia, Albania, and Greece. Here, too, archaeological data are in agreement, and numerous sites from this region have been found with occupations dating to the LGM.

The radical climatic shift in the LGM was not without dire consequences, however. The Cro-Magnons appear to have fared reasonably well during the LGM, but Tallavaara and colleagues suggest that the population of Europe significantly decreased during the coldest times, from about 330,000 people at 30 ka BP to around 130,000 people at 21 ka BP. Some of this may be attributed to outmigration from Europe into warmer climes (for example, into the Levant), but it almost certainly reflects localized extinction of many Cro-Magnon groups. The number of people in Europe does appear to have rebounded during a subsequent warming phase known as Greenland Interstadial 1 (about 14,600 years ago).[3] According to Tallavaara and colleagues, by 13,000 years ago there were perhaps as many as 410,000 people living in Europe.

The largely treeless Europe of the Late Pleistocene is an environment without a modern analog. It is superficially similar to the treeless North Slope of Alaska occupied by the caribou-hunting Nunamiut (chapter 10). However, due to its lower latitude, the cold European plains would have had a longer growing season and at least some hours of daylight even in the depths of winter. In fact, it is not unreasonable to argue that the LGM European plains were more like a cold climate version of the African savanna.

The African savanna is unique among today's terrestrial environments in that the biomass of its plant community is relatively depressed but nonetheless supports a high mammalian biomass, including megafauna such as elephants and rhinoceroses.[4] Both elephants and rhinos had cold-adapted cousins roaming the LGM European plain, and we know the Cro-Magnons saw them if for no other reason than that they created art depicting them! As previously discussed, this cornucopia of animals served not only as food but as a source for tools, building materials, clothing, art objects, and even fuel. In terms of animal resources, the cold steppe and steppe tundra of Europe was a rich environment provided people had sufficient means to deal with the cold.

CULTURAL BUFFERING FROM COLD

Homo sapiens is the most widespread mammalian species on the planet. We are found in frigid Arctic (and Antarctic) environments, and on hypoxic high mountains and plateaus, arid deserts, hot and cold grassy plains, tropical, temperate, and boreal forests, and small atolls in the middle of the Pacific. Although there is evidence for biological adaptations to environmental stressors in humans (e.g., EPAS1, a hypoxia pathway gene that affects hemoglobin at high altitude and has undergone positive selection in native Tibetans[5]), the overwhelming factors affecting our ability to survive in extreme environmental conditions are technological means of ameliorating these stresses—what is known as "cultural buffering." The Cro-Magnons would have been no exception to this rule when it came to dealing with the cold of glacial Europe.

The most important means to deal with cold is through the controlled use of fire. How long had our ancestors been capable of this? There is evidence (albeit controversial) for the controlled use of fire by hominins in Africa around a million years ago. The earliest evidence for the controlled use of fire that is accepted by most paleoanthropologists is about 400,000 to 600,000 years ago at Zhoukoudian in China, the site famous for its now lost *H. erectus* remains. Lewis Binford went to his grave skeptical of the Zhoukoudian evidence, but I cannot imagine Pleistocene hominins living at Zhoukoudian (which at 39° N latitude is the same latitude as Kansas City or Philadelphia) without the controlled use of fire. I wouldn't want to live through a winter outdoors here in New Orleans (30° N latitude) without fire, much less one in Philly or KC! And I would make the same argument regarding clothing.[6]

We also know that the Neandertals enjoyed the controlled use of fire. The vast majority of hearths made by Neandertals are little more than a flat spot, or a slight "bowl," upon which a fire was kindled. In contrast, most Cro-Magnon hearths lie within dug-out depressions (*cuvettes* in French). Flat hearths tend to burn at a higher temperature than those within a depression or bowl, but fires built within bowls are more efficient and require less fuel. One problem with interpreting this pattern as purely functional has been pointed out by the British archaeologist Sophie Jørgensen-Rideout, who notes that the more times a hearth is reused the more likely it is to be more deeply dug out because cleaning the hearth for reuse will tend to remove underlying fire-cracked rock.[7] Thus it could be that most of the Neandertal hearths we see were only short duration fires.

As global temperatures began to drop in the millennia prior to the LGM, Cro-Magnons introduced improvements to fireplaces that made them better at retaining heat. First, at multiple sites across Europe, they began filling the base of the hearth with river cobbles. At Abri Pataud, Jørgensen-Rideout sees this behavior emerging in the late Aurignacian. These stone linings probably served multiple purposes. The first is so-called heat banking, in which the stones absorb heat from the fire and then radiate that heat for hours after the fire has gone out. This is a much more efficient means of heating a space overnight than constantly feeding a fire. The cobbles in question frequently crack from repeated heating and cooling; archaeologists refer to these broken stones as fire-cracked rocks. Our team found many of these within the hearths at Alecrim.

Heated cobbles could also be used for what Jørgensen-Rideout calls "hot rock cookery," which can involve either dry or wet rocks. The dry method is simply using heated rocks to cook packages of food, much like a Dutch oven. The wet process is called "stone boiling," in which the heated cobbles are placed inside a water-filled vessel usually made of the stomach, bladder, or tightly stitched hide of a large mammal. This vessel may be placed within a dug-out pit to give it some structure, and pits adjacent to hearths have been found in sites across Europe. Hot stones added to the vessel bring the water within to a boil, and then bones with or without meat on them are placed within it. This allows one to make either a soup or a stew, and it could also be a method for skimming rendered fat (grease) off the water's surface for use as food or fuel. The archaeologist Rebecca Wragg Sykes notes that roasting game over an open fire is a "caveman cliché," but it would have been horribly wasteful in terms of losing the calorie-rich fat and grease from the animal.[8] Making soup or stew was almost certainly the more frequently utilized option. Note, too, that it is a truism in bioanthropology that humans in cold regions tend to consume more calories than humans living in warmer regions—if for no other reason than to help keep their bodies warm.

Another innovation frequently seen in the hearths of the Cro-Magnons is digging tunnels or trenches to more effectively channel oxygen to the fire, allowing it to burn at a higher temperature. With the exception of some extraordinary hearths such as those at Abric Romaní in Catalonia, this innovation is seldom seen among the Neandertals.[9] But it becomes ubiquitous among the Cro-Magnons during and following the LGM, and it is known from even earlier Upper Paleolithic contexts in southwestern France. Also, the living spaces within caves in multiple Upper Paleolithic locales suggest that low artificial stone walls were constructed around a single hearth or a group of hearths. These walls were probably topped with wooden and animal skin superstructures. Jørgensen-Rideout notes that the presence of *pierres anneaux* ("ring stones," a term attributed to Randall White) at several Upper Paleolithic sites in France is consistent with this idea. These stones are limestone blocks into which a hole has been drilled that are usually found along the shelter's drip line (i.e., its outer edge). It is likely that a rope or cord was suspended from these holes and that animal hides were draped over them. Hanging hides, both at the outer edge of the cave and within it, would have blocked the cold outside air from getting in, and it would have helped retain the heat around the hearth(s) as well.

Clothing is another means by which humans culturally buffer themselves against the cold, but because clothing is made of perishable material the earliest clothes are unlikely to be preserved in the archaeological record. However, the phylogenetic split of two morphologically and behaviorally different populations of the human louse (*Pediculus humanus*) into head lice and body lice—the former of which are limited to head hair and the latter which exclusively lay their eggs in clothing—is estimated to have occurred by at least 83,000 and perhaps as early as 170,000 years ago. This suggests that *H. sapiens* may have already been wearing clothes before we expanded beyond Africa.[10] Given the latitudes at which they lived, Neandertals would probably have draped hides or skins over themselves (they had stone scrapers and bone tools called *lissoirs* for working hides), even though this form of clothing would leave no trace. But what about fitted, tailored clothes? The first eyed needle, dating to about 50,000 years ago and made from a bird bone, was recovered from Denisova Cave in 2016.[11] In Europe eyed needles only become ubiquitous in the Gravettian, in the few millennia just prior to the LGM, a time during which the wearing of tailored clothing would have been a near-necessity for survival.

The animals used for the manufacture of clothing may have also shifted during the LGM. We have every reason to believe that even during the LGM the Cro-Magnons in southwestern France continued to make clothes from the reindeer that were so central to their diet. However, just prior to the LGM in this region, and indeed across Europe, we begin to see mustelids (members of the weasel family) and foxes showing up in archaeological levels. Sometimes the presence of these animals is due to their having dug burrows into those levels. However, in many cases these animals' skeletons show cut marks or are complete skeletons sans paws or are solely represented by their paw skeletons, which is strongly suggestive of their having been exploited for their skins. Most mustelids are small, but in winter they have thick fur coats that would make especially warm clothing if the skins of multiple animals were tightly sewn together with eyed needles. In a study I published in 1998, data from the ethnographic present suggest that the elusive mustelids tend to be trapped, often by women, rather than hunted via ambush, pursuit, or encounter strategies.[12] Unfortunately, simple snares or deadfall traps would leave no trace in the archaeological record.

Footwear would also be desirable during the Last Glacial Maximum. In 2005 Trinkaus looked for evidence of the use of stiff footwear in the Paleolithic.[13]

Within the caves of western Europe, he noted that the vast majority of footprints are of people walking barefoot. One notable exception, a footprint from Grotte de Fontanet in France, is said to have been made by someone wearing a soft, flexible, moccasin-like shoe. Yet caves tend to have the same mild temperature year-round, and just like many people do around the world today, it is likely that shoes were removed once they were inside the living area. Trinkaus notes, too, that multiple burials from the site of Sunghir in Russia, dating just prior to the LGM, include beads about their feet suggesting they were wearing some type of foot covering.

How would one be able to tell if shoes were habitually worn outdoors? Trinkaus examined the bones of the lesser toes of traditionally shod versus unshod humans and found that groups who went without shoes, or who wear flexible soft footwear such as moccasins, tend to have more muscular lesser toe bones than those who wear stiff-soled shoes. He then turned to the paleontological record of Europe and found that Neandertals tended to have strong lesser toe bones, indicating that they either went barefoot or wore flexible foot coverings. In contrast, among humans from just prior to the LGM (Abri Pataud, Barma Grande, Barma del Caviglione, Cro-Magnon, Dolní Věstonice, Paglicci, Předmostí and Veneri), the lesser toe bones are much weaker, indicating that they wore more stiff-soled footwear.

Although they abandoned northern Europe, the Cro-Magnons from just prior to and during the LGM had myriad cultural means for dealing with extreme cold. This came at a steep price, however. The number of people living in Europe plummeted, and there is a genetic discontinuity between the people living in Europe prior to the LGM and those who came later (chapter 6). Some Cro-Magnons did manage to survive the cold by living in the continent's southern refugia. How their descendants would deal with the onset of the Holocene, and the host of problems it brought about, is their final story.

THE PLEISTOCENE SPUTTERS OUT

Figure 12.3 shows the change in $\delta^{18}O$ (a proxy for temperature) over the past 20,000 years, as reflected in the Greenland ice core. The raw data are shown, with a spline line fitted to the data to show their general trend. Once the Last Glacial Maximum ends, these high-resolution data indicate that the temperature does not show a steady rise in the waning millennia of the LGM. Instead the climate goes through a series of rather intense oscillations, moving between colder and warmer (albeit still glacial) temperatures on the scale of centuries rather than

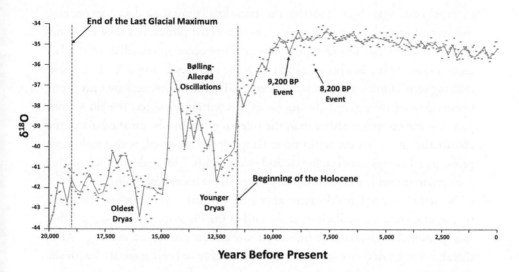

FIGURE 12.3 $\delta^{18}O$ data from the Greenland Ice Core, showing its change across the last 20,000 years; spline line is fit to the data.

Source: Data from NOAA, https://www.ncei.noaa.gov/access/paleo-search/study/17796, accessed August 21, 2021. Plot by author.

millennia. The exact cause of these changes is unknown, but some time around 16,000 years ago, there is a Heinrich event in which huge chunks of ice are discharged into the Atlantic, which could in part be responsible for the Oldest Dryas Stadial.[14] This cold phase, associated with the Magdalenian industry, lasts until about 14,600 years ago, when there is a big spike in temperature marking the beginning of the Bølling Interstadial (Greenland Interstadial 1). Of course, with these oscillations the temperature doesn't remain warm; instead we see a sawtooth pattern of geologically short-lived colder stadials and warmer interstadials. The last of these brief interstadials is called the Allerød Interstadial, during which forests expanded north across Europe.[15] The Allerød ends with the return of cold climate in what is known as the Younger Dryas Stadial about 12,900 years ago.

The Younger Dryas is the last cold phase of the Pleistocene, after which temperatures rise and plateau off in the Holocene. Note that the sawtooth pattern of temperature change is still seen in the Holocene, but the amplitude of the temperature shifts is much lower. There are two brief (less than a few centuries) climatic cool downs of note after the establishment of warmer temperatures in the early Holocene, both of which may have been the result of Heinrich events. The first occurred about 9,200 years ago, and the second more extreme cold phase occurred 1,000 years later (note that the spline line I fit to these data does not take this second cool phase into account). These are cold phases, but they are none-theless warmer than almost all the "warm" Pleistocene interstadials of the last 20,000 years. What has become clear with the high-resolution $\delta^{18}O$ data is that the long-term climatic stability we have enjoyed in the Holocene, with only subtle tweaking of a balmy global climate (as of this writing ice sheets remain at both poles), is the exception rather than the rule. In other words, for the bulk of the Pleistocene, and even deeper in time, temperatures seesawed, with much more pronounced swings between hotter and cold periods. This makes contemplating our current humanly induced climate change all the more frightening.

Nonetheless, the Cro-Magnons appeared to weather (no pun intended) the first several climatic oscillations at the end of the Pleistocene; data suggest that their population continued to increase in size until about 12,000 years ago. After that, however, a decrease in population size may have been associated with the Younger Dryas Stadial. Epipaleolithic sites dating to this phase tend to be smaller, and therefore likely were occupied by fewer people, than the Upper Paleolithic sites that immediately preceded them.

After the end of the Younger Dryas, warmer temperatures associated with the onset of the Holocene brought their own challenges. Perhaps the greatest of these was the reforestation of Europe. There is a reason European fairy tales treat forests as places one should avoid; they are not an easy place for humans to make a living. Forests are an ecosystem in which most of the primary productivity, the production of organic material by plants using energy from the sun and nutrients from the soil, is put into forms that are dietarily inaccessible to mammals. Recall that the European steppe tundra was an ecosystem that was home to multitudes of large gregarious mammals—veritable meat on the hoof that could easily sus-tain large groups of Cro-Magnon hunters. In contrast, the vast majority of bio-mass in a forest is locked up in the wood of tree trunks and branches—structures that most animals simply cannot eat.[16] Reforestation therefore brought about mass extinctions, the timing of which remains somewhat uncertain. Mammoths, woolly rhinoceroses, cave bears, cave hyenas, and cave lions all become extinct, and the European bison population crashes; ultimately, forest bison (wisents) evolve in response to the environmental shift. Reindeer survive in the far north of the continent, but in the south of France and elsewhere they disappear.

Adding insult to injury, mammals in forests tend to be smaller than their cous-ins on the plains. Arboreal (tree-living) mammals tend to be smaller because of the danger of falls, and due to the relatively small package size of edible resources such as nuts. Limited resource availability may also select for smaller bodies. Tem-perate and cold-climate forest mammals are also less gregarious than their cousins on the plains because edible resources such as berries or nuts are more patchily distributed. This means that groups traveling together may not find enough food in a particular patch to allow all group members to eat their fill. As a result, foraging becomes a more solitary activity; for some mammalian species living in forests, the biggest groups are solitary mothers with their dependent offspring.

Hunting small, solitary animals in a forest is no easy task. As the large, gregarious species on which Cro-Magnons had relied in the Pleistocene gradually disappeared, either globally (e.g., mammoths) or locally (e.g., reindeer), the Epipaleolithic descendants of the Cro-Magnons were forced to make adjustments. At first these adjustments were relatively minor, and people largely continued to follow their ancestors' Pleistocene lifeways. In fact, the European Epipaleolithic spans the Pleistocene-Holocene transition, and Epipaleolithic tools are generally slightly smaller versions of the Upper Paleolithic tools made by their Cro-Magnon ancestors.[17] Like the Cro-Magnons before them, Epipaleolithic people continue to hunt red deer (*Cervus elaphus*), roe deer (*Capreolus capreolus*), and wild boars (*Sus scrofa*). What these three species have in common is that they are all extremely adaptable. Red deer can eat almost anything and live pretty much anywhere across Europe. Roe deer fill an ecological niche in Europe similar to that of the widespread New World genus *Odocoileus* (white-tailed and mule deer) and are found in a host of different habitats. Pigs are omnivorous and are famously capable of exploiting a variety of ecosystems as well.

Toward the end of the Upper Paleolithic and Epipaleolithic humans began to expand their diet breadth to include foods that in the past they seldom ate or even ignored. In a classic 1980 *Scientific American* article, Lawrence Straus and colleagues demonstrated this phenomenon for the northern Spanish site of La Riera.[18] Measuring the shell size of limpets (marine snails) in the cave over the last 40,000 years, they found that for most of the site's occupation, shell size rises and falls, perhaps following climate change, but hovers about a long-term average. Late in the Pleistocene, however, during the Magdalenian, it falls below its previous minimum and continues to fall. By the end of the Pleistocene and earliest Holocene, shell size has plummeted to previously unseen lows due to overexploitation by the Magdalenian and later Epipaleolithic inhabitants of the cave. This finding is mirrored in Straus and colleagues' tallying of the ratio of limpet shell mass relative to mammal bone mass for each of the levels. For most of the site's history, mammal bone outweighs the shells, but for the late Magdalenian and Epipaleolithic, the shell mass is 2 to 4.5 times the mass of the mammalian bones from the same levels!

Led by archaeologists Telmo Pereira and Vânia Carvalho, we are currently finding a similar pattern at the site of Abrigo do Poço (Rock Shelter of the Well) in the Chitas Valley of central Portugal. This site is about 26 kilometers (16 miles) from the Atlantic coast at an altitude of around 100 meters (328 feet) above sea level; it would have been slightly farther from the coast in the Pleistocene and early Holocene. Located in the small valley of a seasonal stream, Poço is in a strategic location for hunter-gatherers, lying along a straight-line route from the River Lis to the eastern mountains. During the LGM, Poço appears to have been used almost exclusively as a place to extract high-quality flint from an outcropping that is now exhausted, but in the Epipaleolithic it was more intensely occupied. Radiocarbon dates from the Epipaleolithic occupation give an age corresponding to the 9.2 ka BP cold event. As was the case at La Riera some 480 kilometers (305 miles) to the northeast, the fauna for the Epipaleolithic occupation at Poço is dominated by brackish and saltwater shellfish, which may have been cooked over a fire. Given the distance from the coast, these species were without a doubt brought here by people.

The heavy exploitation of shellfish in Europe continued to grow after the Epipaleolithic. During the Mesolithic, which began about 9,500 years ago, we find huge accumulations of shell, known as middens, in Spain, Portugal, Italy, Brittany, Ireland, Denmark, and the Baltic Sea coast. This is not to say that Mesolithic people subsisted solely on shellfish; it's fairer to say that they were efficient at exploiting anything edible, including red deer, roe deer, moose, aurochs, boars, and rabbits. Many of them gathered hazelnuts and acorns, which they stored over the winter. Others focused efforts on salmon or a host of other fish. Some may have even dabbled at domesticating native plants. In the end, faced with uncertainty after climate change quashed their Pleistocene way of life, the descendants of the Cro-Magnons adapted to their circumstances. When full-scale agriculture was introduced to Europe from the Near East, we also know from ancient DNA that many Mesolithic people readily adopted it. But that is another story, and one beyond the scope of this book.

I want to end this book on a positive note. The optimistic side of me thinks that the lesson of the Cro-Magnons' story is that we humans are a resilient species. Our species' ability to quickly adopt new technologies and new ways of doing things within the lifetime of an individual has divorced us from many selective pressures and made us the predominant mammalian species on the planet. However, the climatic crises the Cro-Magnons endured were not without cost, with perhaps half their population dying. As I write these words just after the release of the sixth assessment report from the Intergovernmental Panel on Climate Change, I remain hopeful that humanity will develop technological means to prevent the worsening of our current climate crisis. At the same time, I realize that time is running out. We've weathered climate change before; my fingers are crossed that we can do it again, and this time I hope we do it without the catastrophic losses.

Acknowledgments

Writing this book has brought back a flood of memories of my 1994 dissertation data collection trip, funded in part by the National Science Foundation (#SBR-9321339) and the Leakey Foundation. In that dissertation I thanked the curators and others who helped me in my travels, but there are a few people I wish to thank here again. Valérie and Ericka Galichon opened their home in Paris to me, for which I am eternally grateful. Enzo Formicola was invaluable with logistics in Italy, and he and I had innumerable great times together in Dolní Věstonice, Pisa, and Massa. The ever-gracious Bernard Vandermeersch moved mountains for me in France; for this I remain in his debt. Jiří Svoboda warmly welcomed me to Dolní Věstonice in 1994, 1999, and 2000; one day soon I hope to take that scenic walk with him to Pavlov for a fine meal. Dan Lieberman was a gracious host in Cambridge, Massachusetts; there I experienced many a "Star Wars" moment in which he would speak English to locals, they would then respond in an indecipherable language, to which his English responses were apparently understood. I also thank my Aunt Gayle and Uncle Wayne, who warmly welcomed me into their Fort Worth home and gave me use of one of their cars so I could make the daily commute to Dallas to study the Jebel Sahaba skeletons.

Following my work on the Lagar Velho child, João Zilhão invited me and my then-doctoral student Vance Hutchinson to excavate in Portugal's Lapedo Valley, where we began the initial exploration of Abrigo do Alecrim. Thank you, Vance, for making that happen, and for constantly making me laugh. Thanks, too, to Francisco Almeida, Telmo Pereira, and Linda Tuero Lindsley for all you did in aid of that work, and to Diego Angelucci, Catia Araújo, Susana Carvalho, Vânia Carvalho, Katina Lillios, and João Zilhão for many enjoyable evenings during those field seasons. I also thank the Tulane University Research Enhancement Fund for helping fund those excavations, and the Tulane University School of Liberal Arts for help with image copyright costs.

Although it gets only passing mention in this book, my collaborative experiences with the *Australopithecus sediba* and *Homo naledi* teams have been among the most intellectually stimulating of my career. I thank Lee Berger, Steve Churchill, Darryl de Ruiter, John Hawks, and Peter Schmid for inviting me to work on these projects, and fellow team members Kristian Carlson, Alex Claxton, Lucas Delezene, Jerry DeSilva, Marina Elliot, Lukas Friedl, Heather Garvin, Tea Jashashvili, Marisa Macias, Damiano Marchi, Sandra Mathews, Tawnee Sparling, and Chris Walker for their thoughtful work in crafting analyses of these important fossil specimens.

In recent years I have been a member of two of the best archaeological teams with which one could ever have the pleasure of working. First, thanks to Telmo Pereira for inviting me to be a part of his EcoPLis project, funded in part by Wenner-Gren (GR 9296) and Portugal's Fundação para a Ciência e a Tecnologia (project IF/01075/2013), for which I excavated at the sites of Abrigo do Poço and Abrigo de Moira with Sandra Assis, Cleia Detry, Marina Évora, João Marreiros, Patrícia Monteiro, David Nora, Eduardo Paixão, and Carlos Simões (and many other volunteers). Thanks for all the fun—the "Italian Boy Scout leader" and "prehistoric boot" stories will have to wait for another book! *Muito obrigado, todos*!

Second, thanks to Bruno Maureille, whom I have known since we were graduate students, for inviting me to take part in new investigations at Regourdou. This work was funded in part by the Agence Nationale de la Recherche (ANR-10-LABX-52), the Région Nouvelle Aquitaine, and the Louisiana Board of Regents LEQSF (2015–18)-RD-A-22. Working with Jaroslav Brůžek, Emmanuel Discamps, Christine Couture-Veschambre, Asier Gómez Olivencia, Christelle Lahaye, François Lacrampe-Cuyaubère, Erwin Le Gueut, Stéphan Madelaine, Xavier Muth, Maxime Pelletier, Rebeka Rmoutilová, Aurélien Royer (who invited me to work on his team at La Balutie), Jean-Pierre Texier, Alain Turq, and many others these past few years has been an amazing journey. Thanks, too, to Michèle Constant and Jean-Charles Cournil, proprietors of the site of Regourdou, for their warm hospitality. *Tout le monde, j'ai l'espoir de vous voir avant longtemps, et je vous donne ma parole qu'un jour avenir, on va avoir un baril d'agrément, nous-autres*!

I thank my editor at Columbia University Press, Miranda Martin, for her immediate and ultimately long-term interest in this manuscript. For a first-time author, having someone happy to shepherd your project is a great gift.

As I was writing this book, many friends and colleagues provided information and advice. I thank Steve Churchill, Jerry DeSilva, Bob Franciscus, Jonathan Haws, James Kidder, Jeff Long, Adeline Masquelier, Bruno Maureille, Andy McDowell, Ozzie Pearson, Fred Smith, Chris Stringer, Erik Trinkaus, Allison Truitt, Sébastien Villotte, and Tim Weaver for promptly and cheerfully answering my out-of-the-blue questions. Thanks to Thomas Bayer and Jason Nesbitt for constantly forwarding relevant articles to me. Special thanks go to Ian Tattersall for providing expert advice on publishing a paleoanthropological book for popular consumption, to Franco Marcantonio for reading and critiquing the paleoclimatology primer, and to Telmo Pereira for critically reviewing the tables from chapter 2. I thank Sophie Jørgensen-Rideout for sharing their informative master's thesis with me, and Enzo Formicola, Chuck Hilton, Jiří Svoboda, Erik Trinkaus, and the Anthropos Institute of the Moravian Museum for allowing me to use their photographs for free. I also thank the two prepublication reviewers of the book, John Hoffecker and John Shea, both of whom chose to reveal their identities. They offered sage advice, and the book is better for their having read it. Mistakes of course remain; for those I take full responsibility. Thanks, too, to John Shea for offering the free use of his artifact drawings, which enhance the book's quality.

Thanks to Erik Trinkaus for being such a close mentor and friend. I hope you like this book, Erik—there's a lot of you in it, and I don't just mean the parts where I talk about your work. Thanks also to my former doctoral students: Lukas Friedl, Joanna Gautney, Vance Hutchinson, Whitney Karriger, Boryana Kasabova, and Brian Pearson. Each of you has made me criticize my own work and question my assumptions, and all of you have made me a better scholar. I hope I have made as much of a positive impact on your lives as you have made on mine.

Finally, I thank my family. My mother and father long supported my desire to be a paleoanthropologist. Dad always wanted me to write a book, so I'm sad he's not here to read it (and Mom, I know you'll love it). I hope my in-laws, Jim and Ann Morgan, enjoy this book—thanks for making me a part of your family for more than two decades! For my adult children, James and Christine, growing up with a dad who disappears for months at a time to go digging across the Atlantic could not have been easy, so thank you for all the love and patience you've shown me over the years. I'm extremely proud of you both. And last, but certainly not least, to my wife, Kathleen Morgan, thank you for always supporting me, for always being there when I need you, and for taking care of everything during my long absences. Thank you, too, for turning your keen scientific and editorial eye to several of these chapters, which helped me avoid multiple pitfalls as I wrote the others. This one is for you, my love.

Notes

INTRODUCTION

1. Broca documented the presence of four adults and one child. Thibeault and Villotte argue that only three adults are represented among the postcranial (below the head) remains, and Partiot et al. demonstrate there are at least three newborns and one infant represented among the children! Paul Broca, "Sur les crânes et ossements des Eyzies," *Bulletins de la Société d'Anthropologie de Paris* 3 (1868): 350–92; Adrien Thibeault and Sébastien Villotte, "Disentangling Cro-Magnon: A Multiproxy Approach to Reassociate Lower Limb Skeletal Remains and to Determine the Biological Profiles of the Adult Individuals," *Journal of Archaeological Science: Reports* 21 (2018): 76–86; Caroline Partiot, Erik Trinkaus, Christopher J. Knüsel, and Sébastien Villotte, "The Cro-Magnon Babies: Morphology and Mortuary Implications of the Cro-Magnon Immature Remains," *Journal of Archaeological Science: Reports* 30 (2020): 102257.

2. The technical term for rock shelter in French is *abri sous-roche*, but even French scholars tend to abbreviate this to *abri*. A rock shelter is what many of us would refer to as a cave because they tend to have cavelike entrances into a cliff or bluff, frequently, but not always, at the cliff's base. However, unlike true caves, rock shelters are not deep and do not have subterranean passages. In French, caves are referred to as *grottes* or *cavernes*.

3. Occitan is a Romance language with multiple dialects that was once widely spoken across southern France, parts of Catalonia, and a small portion of Italy bordering southern France. Although it had fallen into disuse over most of its former area, the language is currently enjoying a bit of a renaissance.

4. Louis Lartet, "Une sépulture des troglodytes du Périgord," *Bulletins de la Société d'Anthropologie de Paris* t. 3 (1868): 335–49.

5. Stringer argues the earliest fossil *Homo sapiens* specimens in Africa, like Jebel Irhoud 1, are not anatomically modern because they retain many ancestral ("primitive") characters (chapter 4). Chris Stringer, "The Origin and Evolution of *Homo sapiens*," *Philosophical Transactions of the Royal Society B* 371 (2016): 20150237.

6. Here and throughout the book I adopt the spelling "Neandertal." When the first recognized Neandertal was discovered in the Neander Valley in 1856, the German word for valley was *thal*, hence the name "Neanderthal." In the early twentieth century, however, Germans changed their orthography such that spellings more nearly matched pronunciation, and the spelling of the word *thal*

was changed to *tal*. If the Germans now call them "Neandertals," my feeling is so should we! Linnaean taxonomic names are much more resistant to change; thus we retain the spelling of *Homo neanderthalensis*.

7. The Pleistocene is a geological epoch characterized by multiple glacial ("ice age") phases separated by warmer interglacial episodes. It began 2.58 million years ago and ended 11,700 years ago with the onset of the current interglacial phase known as the Holocene. The Pleistocene is divided into three broad phases: (1) the Early Pleistocene, from 2.58 million years ago to 774,000 years ago; (2) the Middle Pleistocene, from 774,000 years ago to 129,000 years ago; and (3) the Late Pleistocene, from 129,000 years ago to 11,700 years ago. Most fossil hominins discussed in this book are dated to the Late Pleistocene.

8. Charles Lyell, *Principles of Geology*, 6th ed. (London: John Murray, 1840).

9. We spend some time wrestling with what species are in chapter 6.

10. Ontology is the realm of philosophy associated with the nature of being; i.e., the essence of things— thus *ontologically* is broadly similar to the adverb *essentially*.

11. Charles Darwin, *On the Origin of Species by Means of Natural Selection* (London: John Murray, 1859), 55.

12. Lamarck thought parents could pass onto their offspring traits they had acquired during their lifetimes. Were this the case, one might imagine that the children of Arnold Schwarzenegger would be born with huge muscles.

13. William Sollas was the first to recognize that the Paviland 1 remains were those of a young adult male. William J. Sollas, "Paviland Cave: An Aurignacian Station in Wales," *Journal of the Royal Anthropological Institute* 43 (1913): 325–74.

14. Édouard was the father of Louis Lartet (1840–1899). Louis was a paleontologist like his father and was put in charge of the Cro-Magnon site in 1868. He wrote what was for the time an excellent description of the site (as much as could be salvaged from the damage done by the railway workers) primarily from a geological perspective, at the relatively young age of twenty-eight (chapter 1).

15. I hesitated to cite this work by Morant because it was published in a eugenics journal. Eugenicists believed that certain groups and certain types of people should not be allowed to reproduce, a morally repugnant position. Unfortunately, eugenicists were influential in the United States in the decades just prior to World War II, and their influence extended to Hitler and the Third Reich in Germany (chapter 9). In the end, despite my discomfort, I decided the information Morant provides on Paleolithic sites and the skeletons found within them is useful, at least from a historical perspective. G. M. Morant, "Studies of Palaeolithic Man IV. A Biometric Study of the Upper Palaeolithic Skulls of Europe and Their Relationships to Earlier and Later Types," *Annals of Eugenics* 4 (1930): 109–214.

16. Erik Trinkaus and Pat Shipman, *The Neandertals: Changing the Image of Mankind* (New York: Knopf, 1992).

1. DISCOVERY

1. The Périgord is traditionally divided into four areas, designated by colors. The Black Périgord, in which the Cro-Magnon shelter lies, is in the southeastern portion of the region, roughly corresponding to the area drained by the Vézère River and the Dordogne River upstream from where the Vézère joins it. It is called black because of its dense forests. The other areas are the Purple Périgord in the southwest (purple for the wine vinted in and around Bergerac), the White Périgord in the central area, named for the color of its limestone, and the Green Périgord in the north, named for its lush vegetation.

2. Louis Lartet, "Une sépulture des troglodytes du Périgord," *Bulletins de la Société d'Anthropologie de Paris* t. 3 (1868): 335–49.

3. Dominique Henry-Gambier, "Les fossiles de Cro-Magnon (Les Eyzies-de-Tayac, Dordogne) Nouvelles données sur leur position chronologique et leur attribution culturelle." *Bulletins et Mémoires de la Société d'Anthropologie de Paris* 14 (2002): 89–112.

4. Lartet, "Une sépulture des troglodytes du Périgord."

5. Lartet, "Une sépulture des troglodytes du Périgord."

6. Lartet, "Une sépulture des troglodytes du Périgord."

7. An index fossil is a fossil of known age or paleontological context that can be used to date or contextualize other deposits in which it is found.

8. This nonvernacular construction may strike the reader as odd, but when writing about or discussing a group of organisms being given a certain species name, a taxonomist (someone who practices taxonomy, the science of naming and classifying organisms) will say the organisms are "referred to Species X."

9. Blades are elongated flakes; one simple definition is that a blade is at least twice as long as it is wide.

10. Harvey M. Bricker, "The Provenience of Flint Used for the Manufacture of Tools," in *Excavation of the Abri Pataud (Les Eyzies, Dordogne)*, ed. Hallam L. Movius Jr. (Cambridge, MA: American School of Prehistoric Research, 1975), 194–98. In addition to teaching me much about the European Paleolithic, Harvey was my faculty mentor when I was a junior faculty member at Tulane. I will always be grateful for the steadfast support and guidance he gave me while I was navigating the tenure process.

11. Paul Broca, "Sur les crânes et ossements des Eyzies," *Bulletins de la Société d'Anthropologie de Paris* 3 (1868): 350–92. After extensive analysis, Thibeault and Villotte agree that this femur is associated with the Cro-Magnon 1 cranium. See Adrien Thibeault and Sébastien Villotte, "Disentangling Cro-Magnon: A Multiproxy Approach to Reassociate Lower Limb Skeletal Remains and to Determine the Biological Profiles of the Adult Individuals," *Journal of Archaeological Science: Reports* 21 (2018): 76–86.

12. Jean Dastugue, "Pathologie des hommes fossiles de l'Abri de Cro-Magnon," *L'Anthropologie* 71 (1967): 479–92.

13. Henry-Gambier, "Les fossiles de Cro-Magnon (Les Eyzies-de-Tayac, Dordogne)."

14. This is my own translation of a quotation from Paul Fischer, "Sur la conchyliologie des cavernes," *Bulletins de la Société d'Anthropologie de Paris* Deuxième série, 11 (1876): 181–85, at 185.

15. In Italian, a cave is usually called a *grotta*. The Italian word for rock shelter is *riparo*; however, in the local dialect in and around Ventimiglia, the word *barma* can refer to either a cave or a rock shelter.

16. Henry de Lumley, ed., *La Grotte du Cavillon sous la falaise des Baousse Rousse, Grimaldi, Vintimille, Italie (Archéologie/Préhistoire)* (Paris: CNRS Éditions, 2016). I'd rather not air too much dirty laundry in this book, but I was denied access to study this specimen in 1994, being told a publication on it was imminent. A further request to study it in 2000 was met with a similar denial. The last such request I made in 2009 went unanswered. Science rests on access to data and replicability, and it is therefore my hope that the next generation of paleoanthropologists will be more open with access to fossils, at least once their discoverers have had a chance to publish on them. Perhaps in this particular case I was a bit too hasty—the specimen was only discovered in 1872 after all!

17. Vincenzo Formicola and Brigitte M. Holt, "Tall Guys and Fat Ladies: Grimaldi's Upper Paleolithic Burials and Figurines in an Historical Perspective," *Journal of Anthropological Sciences* 93 (2015): 1–18.

18. René Verneau, *Les Grottes de Grimaldi (Baoussé-Roussé). Anthropologie* (Monaco: Imprimerie de Monaco, 1906).

19. Arthur J. Evans, "On the Prehistoric Interments of the Balzi Rossi Caves Near Mentone and Their Relation to the Neolithic Cave-Burials of the Finalese," *Journal of the Anthropological Institute of*

Great Britain and Ireland 22 (1893): 286–307. This almost certainly exaggerates the height of this individual (chapter 9). Once in the laboratory, Verneau estimated the individual's stature at about 177 cm (5 ft. 10 in.), although I suspect he was a bit taller than that. BG2 was a big person, but he was not a "giant" by any stretch of the imagination.

20. Verneau, *Les Grottes de Grimaldi (Baoussé-Roussé)*.

21. Formicola and Holt, "Tall Guys and Fat Ladies."

22. Evans, "On the Prehistoric Interments of the Balzi Rossi Caves Near Mentone," 293.

23. Verneau, *Les Grottes de Grimaldi (Baoussé-Roussé)*.

24. Verneau, *Les Grottes de Grimaldi (Baoussé-Roussé)*; Formicola and Holt, "Tall Guys and Fat Ladies."

25. In 1861, Édouard Lartet proposed dividing the Pleistocene into three ages, from earliest to latest: the Cave Bear Age, the Elephant and Rhinoceros Age, and the Reindeer Age. Many subsequent scholars favored a two-age system, combining the Cave Bear and Elephant ages into one. These faunal ages are no longer used, but the Reindeer Age is somewhat equivalent to the Upper Paleolithic, defined by the tool industries used by the Cro-Magnons. Note that in Europe there was a huge technological gulf between the tools of the Upper Paleolithic (Reindeer Age) and those of other time periods. The tool industries of the Upper Paleolithic are defined in chapter 2, and further illuminated in later chapters.

2. ARCHAEOLOGY OF THE ANCIENTS

1. The Clovis culture is named for the New Mexico town closest to its type site of Blackwater Draw, where "Clovis points" were first found in 1929. This culture was widespread—from Washington state and Sonora in the west to Nova Scotia and Florida in the east. It may have been present in Panama and Venezuela. It was relatively short-lived, dating from about 13,200–12,900 cal BP (calibrated years before present). For years it was considered the earliest culture in the Americas, but over the last few decades sites such as Meadowcroft Rockshelter in Pennsylvania and Monte Verde in Chile have shown people were in the Americas thousands of years earlier.

2. This date is not universally accepted. Molecular (genetic) data suggest to many that the human-chimpanzee split occurred about 6 million years ago. I don't want to discount molecular divergence times, as all data with regard to human evolution are valuable, and molecular data certainly proved their worth in the 1960s and 1970s when they corrected the paleontological notion of "Ramapithecus" as a 13-million-year-old hominin. However, almost all molecular dates are rooted in the fossil record, and with fossils we never expect to find the earliest representative of any lineage or species. Therefore, molecular dates, like fossil dates, tend to skew young. In addition, as of this writing, I accept the 7-million-year-old species *Sahelanthropus tchadensis* as hominin. Thus the human-chimpanzee split must be earlier than 7 million years ago (so I chose 8 million). That said, I could be convinced to change my mind with more data. We paleoanthropologists are skeptics, but we try to keep an open mind.

3. The paleoanthropologist Steve Churchill of Duke University tells a funny story from when he was part of an international team excavating the Paleolithic site of Kebara Cave in northern Israel. At some point, two American undergraduate students approached the delightfully affable Ofer Bar-Yosef (1937–2020), the Harvard professor in charge of the dig, complaining that they thought the levels they were digging were Paleolithic. Ofer assured them that they were. The American students then asked why the French students were constantly talking about the Bronze Age. *Bronzage* in French is to get a suntan or sunbathe. The term for Bronze Age in French is *Âge du Bronze*.

4. Sonia Harmand et al., "3.3-Million-Year-Old Stone Tools from Lomekwi 3, West Turkana, Kenya," *Nature* 521 (2015): 310–15.

5. An archaeological industry is a group of technologically similar tool complexes that are spatially and temporally bounded. Dividing the archaeological record into temporally and spatially bounded industries dates back to Gabriel de Mortillet, who defined French Paleolithic industries based on artifact types, especially "type fossils," the presence of which could tell the archaeologist exactly what industry a level represented. Archaeological industries were also rooted in "type sites" for which each industry is named. This system, with extensive modification, is still in use by Paleolithic archaeologists. There is, however, a vocal minority who argue that this methodology has outlived its usefulness. Gabrielle de Mortillet, "Classification des diverses périodes de l'Âge de la pierre," *Congrès International d'Anthropologie et d'Archéologie Préhistorique, Bruxelles* (1871): 432–59.

6. I am aware that by the traditional definition of Europe, most of the Republic of Georgia lies in Asia. That said, Europe's boundaries are an arbitrary abstraction, and Georgia considers itself a European nation.

7. There is some evidence that the eruption of Vesuvius in 79 CE occurred in October, not in August as Pliny the Younger is said to have written. His letters may therefore have been mistranslated. G. Rolandi et al. "The 79 AD Eruption of Somma: The Relationship Between the Date of the Eruption and the Southeast Tephra Dispersion," *Journal of Volcanology and Geothermal Research* 169 (2007): 87–98

8. Rachel E. Wood et al., "Radiocarbon Dating Casts Doubt on the Late Chronology of the Middle to Upper Palaeolithic Transition in Southern Iberia," *Proceedings of the National Academy of Sciences (USA)* 110 (2013): 2781–86.

9. R. M. Jacobi and T. F. G. Higham, "The 'Red Lady' Ages Gracefully: New Ultrafiltration AMS Determinations from Paviland," *Journal of Human Evolution* 55 (2008): 898–907.

10. This stands for calibrated years before present, with "present" being the year 1950.

3. THE ABEL TO OUR CAIN?:
HOMO NEANDERTHALENSIS

1. This was not an original idea of Boule's. In his initial study of the Spy Neandertal remains, Julien Fraipont noted a similar incline of the tibial plateau (the flat joint surface at the bone's proximal, or knee, end). He thus attributed to Neandertals a stooped, bent-kneed gait. The truth is that this declination is an indicator of vigorous activity, and modern humans who are active during childhood and adolescence tend to show this morphology. How does this happen? During growth and development, joints orient themselves to be perpendicular to the predominant forces that act on them. The maximum force the proximal tibia experiences occurs when the knee is slightly flexed; a declination in the tibial plateau places the plateau perpendicular to that force.

2. Chris Stringer, "The Status of *Homo heidelbergensis* (Schoetensack 1908)," *Evolutionary Anthropology* 21 (2012): 101–7.

3. Matthias Meyer et al., "Nuclear DNA Sequences from the Middle Pleistocene Sima de los Huesos Hominins," *Nature* 531 (2016): 504–7.

4. If you are interested in reading about this topic in a much more detailed historical narrative form, see Erik Trinkaus and Pat Shipman, *The Neandertals: Changing the Image of Mankind* (New York: Knopf, 1992).

5. Ralf W. Schmitz et al., "The Neandertal Type Site Revisited: Interdisciplinary Investigations of Skeletal Remains from the Neander Valley, Germany," *Proceedings of the National Academy of Sciences (USA)* 99 (2002): 13342–47.

6. In the introduction, I made a point of saying that even the Germans spell "Neandertal" without the "h"—so why is this museum outside of Düsseldorf called the "Neander*th*al Museum?" Four words: English language cultural imperialism. There, I said it.

7. Schmitz et al., "The Neandertal Type Site Revisited."

8. As someone who teaches undergraduates, I am sad to report that Jean Auel's enormously popular book series that began with her first novel, *The Clan of the Cave Bear* (Bantam, 1980), is no longer the cultural touchstone it once was. Most of my students haven't heard of the book or the 1986 movie made from it. I remain a steadfast defender of Neandertal intelligence, but the notion that when we find cave bear skulls in caves it's because Neandertals were caching them as totems drives me nuts. If you had to guess a place where a *cave* bear would most likely be when it dies, what would you say?

9. Although the Feldhofer Cave no longer exists, through archival research in 1997 Ralf Schmitz and Jürgen Thissen were able to deduce where the sediments from Feldhofer Cave had been dumped. They began a project to re-excavate these sediments and found three additional pieces that articulate with the original Neandertal skeleton, as well as the remains of a second adult Neandertal individual and at least one Neandertal subadult (about eleven to fourteen years old at the time of death). In addition, Mousterian tools were recovered. Schmitz et al., "The Neandertal Type Site Revisited."

10. Schmitz et al., "The Neandertal Type Site Revisited."

11. In his defense, the mandible *is* too thick to be that of a modern human child.

12. By 1887 our poor friend Barma del Caviglione 1 had been languishing in Paris awaiting formal study for fifteen years! Hang in there, *Madame* (or should I say *Signora*?); only 129 years left to go!

13. Rendu and colleagues reported results of recent excavations at La Chapelle-aux-Saints showing compelling evidence that LC1 was intentionally buried. The archaeologist Harold Dibble (who died in 2018) and his colleagues challenged this hypothesis in 2015. Within a year, Rendu et al. had refuted Dibble and colleagues' arguments with additional data. The La Chapelle-aux-Saints 1 skeleton was an intentional interment; I cite these works so you can explore this controversy if you wish. William Rendu et al., "Evidence Supporting an Intentional Neandertal Burial at La Chapelle-aux-Saints," *Proceedings of the National Academy of Sciences (USA)* 111 (2014): 81–86; Harold L. Dibble et al., "A Critical Look at Evidence from La Chapelle-aux-Saints Supporting an Intentional Neandertal Burial," *Journal of Archaeological Science* 53 (2015): 649–57; William Rendu et al., "Let the Dead Speak . . . Comments on Dibble et al.'s Reply to 'Evidence Supporting an Intentional Burial at La Chapelle-aux-Saints,'" *Journal of Archaeological Science* 69 (2016): 12–20.

14. When teeth are no longer present in their sockets, it's almost as if the bone surrounding them (the alveolar bone) loses its will to live and disappears at an alarming rate. It is for this reason that dentures must be continually adjusted to fit onto jaws that get smaller and smaller. As I was told in middle school health class, "only floss the teeth you want to keep."

15. Peyrony did not call the uppermost level "Gravettian," instead referring to it as the "Upper Perigordian," which he linked culturally to the earlier "Lower Perigordian" (what archaeologists today call the "Châtelperronian"). Today's archaeologists do not view a link between cultures separated by 10,000 years as plausible. Peyrony also argued that an Acheulean level at La Ferrassie was under the Mousterian one because hand axes were found in the deepest layer. Hand axes are, however, recovered in many Mousterian assemblages. One of the archaeologist François Bordes's (1919–1981) Mousterian facies, marked by a high frequency of hand axes, is called the "Mousterian of Acheulean Tradition" (abbreviated MTA in French). The oldest layers at La Ferrassie are now considered Mousterian.

16. William King, "The Reputed Fossil Man of Neanderthal," *Quarterly Journal of Science* 1 (1864): 88–97.

4. FOSSIL AND RECENT *HOMO SAPIENS*

1. Epiphyses are separate ossification (bony growth) centers found near the ends of long bones. During growth they are connected to the diaphyses (shafts) of long bones at epiphyseal plates (or lines) composed of hyaline cartilage. As the cartilage in the plate grows, it is replaced by bone at the metaphysis (the end of the diaphysis), causing the bone to increase in length. Skeletal maturity occurs when the cartilage in the plate stops growing following a hormonal signal. The plate is then completely replaced by bone, obliterating the epiphyseal line and making future growth in length of the long bone impossible. (The shaft can, however, continue to thicken in cross-section in response to biomechanical forces.)

2. Zhaoyu Zhu et al., "Hominin Occupation of the Chinese Loess Plateau Since About 2.1 Million Years Ago," *Nature* 559 (2018): 608–12.

3. Daniel Richter et al., "The Age of the Hominin Fossils from Jebel Irhoud, Morocco, and the Origins of the Middle Stone Age," *Nature* 546 (2017): 293–96.

4. 3D morphometric analysis involves taking a series of landmark data on an object of interest (in this case, a cranium) as points measured as three-dimensional distances (X, Y, and Z axes) from a zero point. Generalized Procrustes Analysis (GPA) creates a common centroid and orientation for all the data, size-standardizing them. These size-standardized data are then examined using exploratory analyses such as principal components analysis (PCA), which distills complex multidimensional data into a smaller number of more readily interpreted components. Jean-Jacques Hublin et al., "New Fossils from Jebel Irhoud, Morocco and the Pan-African Origin of *Homo sapiens*," *Nature* 546 (2017): 290–92.

5. Chris Stringer, "The Origin and Evolution of *Homo sapiens*," *Philosophical Transactions of the Royal Society B* 371 (2016): 20150237.

6. Hublin et al., "New Fossils from Jebel Irhoud."

7. Hublin et al., "New Fossils from Jebel Irhoud."

8. Amino acid racemization dating takes advantage of the molecular property of chirality, in which the atoms within molecules of the same substance can be arranged such that they are "handed" forms—nonsuperimposable mirror images of each other. During life, amino acids, the building blocks of the body's proteins, are kept in the left-handed configuration. After death, some molecules alter to the right-handed configuration. Over long periods, an equilibrium is reached, with about half the molecules in the right-handed form and half in the left-handed form switching back and forth. If an eggshell or other durable tissue is not completely fossilized, some of its protein may remain; and if the ratio has not reached equilibrium, the handedness of its amino acids is assessed to provide an absolute date. My wife teaches organic chemistry and uses this fact to help her students understand just how slowly racemization takes place. Paul C. Manega, "Geochronology, Geochemistry, and Isotopic Study of the Plio-Pleistocene Hominid Sites and the Ngorongoro Volcanic Highland in Northern Tanzania" (PhD diss. University of Colorado, 1993).

9. Hublin et al., "New Fossils from Jebel Irhoud."

10. Hublin et al., "New Fossils from Jebel Irhoud."

11. Céline M. Vidal et al., "Age of the Oldest Known *Homo sapiens* from Eastern Africa," *Nature* 601 (2022): 579–83.

12. Michael H. Day and Chris B. Stringer, "A Reconsideration of the Omo Kibish Remains and the *erectus–sapiens* Transition," in *L'*Homo erectus *et la Place de l'Homme de Tautavel Parmi les Hominides Fossiles*, ed. Henry De Lumley (Nice: Centre National de la Recherche Scientifique, 1982), 814–46.

13. J. Desmond Clark et al., "Stratigraphic, Chronological and Behavioural Contexts of Pleistocene *Homo sapiens* from Middle Awash, Ethiopia," *Nature* 423 (2003): 747–52; Vidal et al., "Age of the Oldest Known *Homo sapiens* from Eastern Africa."

14. Tim D. White et al., "Pleistocene *Homo sapiens* from Middle Awash, Ethiopia," *Nature* 423 (2003): 742–47.

15. In zoology, a paratype is an additional specimen used to help define a species, or in this case a sub-species. The holotype of any species carries the species name with it wherever it goes; a paratype can later be removed from the species if it is discovered that it belongs to a different taxon.

16. Israel Hershkovitz et al., "The Earliest Modern Humans Outside Africa," *Science* 359 (2018): 456–59.

17. C. Stringer, R. Grün, H. Schwarcz, and P. Goldberg, "ESR Dates for the Hominid Burial Site of Es Skhul in Israel," *Nature* 338 (1989): 756–58; N. Mercier et al., "Thermoluminescence Date for the Mousterian Burial Site of Es-Skhul, Mt. Carmel," *Journal of Archaeological Science* 20 (1993): 169–74.

18. Hublin et al., "New Fossils from Jebel Irhoud." Given how much of the face is missing, I am surprised they were able to include it in their facial analyses.

19. Hublin et al., "New Fossils from Jebel Irhoud."

20. Katerina Harvati et al., "Apidima Cave Fossils Provide Earliest Evidence of *Homo sapiens* in Eurasia," *Nature* 571 (2019): 500–504.

21. Wu Liu et al., "The Earliest Unequivocally Modern Humans in Southern China," *Nature* 526 (2015): 696–700; Yanjun Cai et al., "The Age of Human Remains and Associated Fauna from Zhiren Cave in Guangxi, Southern China," *Quaternary International* 434 (2017): 84–91.

22. Day and Stringer, "A Reconsideration of the Omo Kibish Remains and the *erectus–sapiens* Transition."

23. William W. Howells, *Cranial Variation in Man. A Study by Multivariate Analysis of Patterns of Differences Among Recent Human Populations*. Papers of the Peabody Museum of Archeology and Ethnology, vol. 67 (Cambridge, MA: Peabody Museum, 1973); Milford H. Wolpoff, "Describing Anatomically Modern *Homo sapiens*. A Distinction Without a Definable Difference," in *Fossil Man—New Facts, New Ideas. Papers in Honor of Jan Jelínek's Life Anniversary*, ed. V. V. Novotný and A. Mizerová (Brno: Anthropos, 1986), 41–53.

24. James H. Kidder, Richard L. Jantz, and Fred H. Smith, "Defining Modern Humans: A Multivari-ate Approach," in *Continuity or Replacement: Controversies in* Homo sapiens *Evolution*, ed. Günter Bräuer and Fred H. Smith (Rotterdam: A. A. Balkema, 1992), 157–77.

25. J. N. Darroch and J. E. Mosimann, "Canonical and Principal Components of Shape," *Biometrika* 72 (1985): 241–52.

26. Isometry is a special case of allometry in which the size of a part of the body changes at exactly the same rate as overall body size. Allometry is the study of changes in shape associated with changes in size (see chapter 7).

5. A PALEONTOLOGICAL PERSPECTIVE ON MODERN HUMAN ORIGINS

1. In 2018, the members of the American Association of Physical Anthropologists voted to change the name of the organization to the American Association of Biological Anthropologists, with the change officially taking place in 2021.

2. A genetic test reveals that 1.5 percent of my genes are from Neandertals, below the global average. I had hoped I would have a high proportion of Neandertal genes—that would make a great news-paper headline: "Researcher who studies Neandertals has highest proportion of Neandertal genes on record." Ours is an imperfect universe.

3. Claparède also volunteered to help Clémence Royer (1830–1902), the woman whom Darwin asked to write the French translation of *On the Origin of Species*, with French translations of biological terms. Royer's translation of Darwin's tome was published in 1862.

4. Gene flow is the exchange of genes (alleles) across demic boundaries (a deme is a population) or across geographical distance. Gene flow is a homogenizing force in evolution because populations that are exchanging genes become more genetically similar to each other.

5. After originally opining that Neandertals were not ancestral to modern humans, by 1939, Keith had revised his view, based on examination of Neandertal remains excavated by Dorothy Garrod in the Mount Carmel Caves.

6. Brian Villmoare et al., "Early *Homo* at 2.8 Ma from Ledi-Geraru, Afar, Ethiopia," *Science* 347 (2015): 1352–55.

7. Cladogenesis is the mode of speciation in which a parental species splits into two or more daughter species. As such, it is ultimately responsible for all the diversity one sees in Darwin's "Tree of Life."

8. Milford H. Wolpoff, Xinzhi Wu, and Alan G. Thorne, "Modern *Homo sapiens* Origins: A General Theory of Hominid Evolution Involving the Fossil Evidence from East Asia," in *The Origins of Modern Humans: A World Survey of the Fossil Evidence*, ed. Fred H. Smith and Frank Spencer (New York: Alan R. Liss, 1984), 411–83.

9. Genetic drift refers to random changes in gene frequencies in a population across generations. These changes are not brought about by selection (i.e., differential reproductive success) but rather are "sampling errors" from one generation to the next in which copies of genes known as alleles are randomly lost or fixed in a population. The impact of drift is much greater in small populations than in large ones.

10. Christopher B. Stringer, Jean-Jacques Hublin, and Bernard Vandermeersch, "The Origins of Anatomically Modern Humans in Western Europe," in *The Origins of Modern Humans: A World Survey of the Fossil Evidence*, ed. Fred H. Smith and Frank Spencer (New York: Alan R. Liss, 1984), 51–135.

11. Günther Bräuer, "A Craniological Approach to the Origin of Anatomically Modern *Homo sapiens* in Africa and Implications for the Appearance of Modern Humans," in *The Origins of Modern Humans: A World Survey of the Fossil Evidence*, ed. Fred H. Smith and Frank Spencer (New York: Alan R. Liss, 1984), 327–410.

12. Fred H. Smith, "Fossil Hominids from the Upper Pleistocene of Central Europe and the Origin of Modern Europeans," in *The Origins of Modern Humans: A World Survey of the Fossil Evidence*, ed. Fred H. Smith and Frank Spencer (New York: Alan R. Liss, 1984), 137–209.

13. René Verneau, *Les Grottes de Grimaldi (Baoussé-Roussé). Anthropologie* (Monaco: Imprimerie de Monaco, 1906).

14. Erik Trinkaus, "Neanderthal Limb Proportions and Cold Adaptation," in *Aspects of Human Evolution*, ed. Christopher B. Stringer (London: Taylor and Francis, 1981), 187–224.

15. Joel A. Allen, "The Influence of Physical Conditions in the Genesis of Species," *Radical Review* 1 (1877): 108–40.

16. The bioanthropologist Christopher Ruff was a member of my doctoral committee. By modeling the human body as a cylinder, he has long argued that differences in bi-iliac breadth have a much greater impact on the body's relative surface area than do changes in body height (or limb length). See Boryana E. Kasabova and Trenton W. Holliday, "New Model for Estimating the Relationship Between Surface Area and Volume in the Human Body Using Skeletal Remains," *American Journal of Physical Anthropology* 156 (2015): 614–24; and Ernst Mayr, *Principles of Systematic Zoology* (New York: McGraw-Hill, 1969).

17. M. A. Serrat, R. M. Williams, and C. E. Farnum, "Temperature Alters Solute Transport in Growth Plate Cartilage Measured by In Vivo Multiphoton Microscopy," *Journal of Applied Physiology* 106 (2009): 2016–25; Maria A. Serrat, "Allen's Rule Revisited: Temperature Influences Bone Elongation During a Critical Period of Postnatal Development," *Anatomical Record* 296 (2013): 1534–45.

18. Derek F. Roberts, *Climate and Human Variability*, 2nd ed. (Menlo Park: Cummings, 1978).

19. Robert G. Franciscus and Trenton W. Holliday, "Hindlimb Skeletal Allometry in Plio-Pleistocene Hominids with Special Reference to AL 288–1 ('Lucy')," *Bulletins et Mémoires de la Société d'Anthropologie de Paris* n.s., 4 (1992): 5–20.

20. Trenton W. Holliday, "Body Proportions in Late Pleistocene Europe and Modern Human Origins," *Journal of Human Evolution* 32 (1997): 423–48.

21. The 95 percent confidence ellipse estimates the position of a group's population mean in two-dimensional space. In other words, there is a 95 percent chance that the group mean lies somewhere within the bounds of the ellipse.

22. Trenton W. Holliday and Anthony B. Falsetti, "A New Method for Discriminating African-American from European-American Skeletons Using Postcranial Osteometrics Reflective of Body Shape," *Journal of Forensic Sciences* 44 (1999): 926–30.

6. THE GENETICS OF MODERN HUMAN ORIGINS

1. You've noticed by now that I talk about food a lot. I'm a New Orleanian; it's what we do.

2. Occasionally some of the sperm's mitochondria are incorporated into the fertilized egg.

3. Rebecca L. Cann, Mark Stoneking, and Allan C. Wilson, "Mitochondrial DNA and Human Evolution," *Nature* 325 (1987): 31–36.

4. David R. Maddison, "African Origin of Human Mitochondrial DNA Reexamined," *Systematic Zoology* 40 (1991): 355–63; David R. Maddison, "The Discovery and Importance of Multiple Island Trees," *Systematic Zoology* 40 (1991): 315–28.

5. Svante Pääbo, "Molecular Cloning of Ancient Egyptian Mummy DNA," *Nature* 314 (1985): 644–45.

6. Matthias Krings et al., "Neandertal DNA Sequences and the Origin of Modern Humans," *Cell* 90 (1997): 19–30.

7. Four different nucleotides are used in the DNA code, abbreviated as A, C, G, and T. When comparing any two individuals' DNA sequences, different nucleotides can occupy the same position. In Krings et al., these pairwise nucleotide differences were tallied up and presented as three Gaussian distributions (bell curves): recent human vs. other recent humans, Neandertal vs. recent humans, and chimpanzee vs. recent humans. The Neandertal–recent human differences bell curve was located almost halfway between the human-to-human differences and chimpanzee-to-human differences bell curves. Krings et al., "Neandertal DNA Sequences and the Origin of Modern Humans."

8. Matthias Krings et al., "A View of Neandertal Genetic Diversity," *Nature Genetics* 26 (2000): 144–46; I. V. Ovchinnikov et al., "Molecular Analysis of Neanderthal DNA from the Northern Caucasus," *Nature* 404 (2000): 490–93.

9. Ralf W. Schmitz et al., "The Neandertal Type Site Revisited: Interdisciplinary Investigations of Skeletal Remains from the Neander Valley, Germany," *Proceedings of the National Academy of Sciences (USA)* 99 (2002): 13342–47.

10. David Serre, "No Evidence of Neandertal mtDNA Contribution to Early Modern Humans," *PLoS Biology* 2 (2004): E57.

11. Johannes Krause et al., "Neanderthals in Central Asia and Siberia," *Nature* 449 (2007): 902–4.

12. Richard E. Green et al., "Analysis of One Million Base Pairs of Neanderthal DNA," *Nature* 444 (2006): 330–36.

13. Richard E. Green et al., "A Draft Sequence of the Neandertal Genome," *Science* 328 (2010): 710–22.

14. Green et al., "A Draft Sequence of the Neandertal Genome."

15. Johannes Krause et al., "The Complete Mitochondrial DNA Genome of an Unknown Hominin from Southern Siberia," *Nature* 464 (2010): 894–97.

16. David Reich et al., "Genetic History of an Archaic Hominin Group from Denisova Cave in Siberia," *Nature* 468 (2010): 1054–59.

17. In 2019, Fahu Chen et al. announced that DNA extracted from a 160,000-year-old partial mandible from Baishiya Karst Cave in Tibet (the so-called Xiahe Mandible) was that of a Denisovan. Unfortunately, there is not enough preserved anatomy in this specimen to distinguish it from other forms of Middle Pleistocene *Homo*. In 2021, Qiang Ji et al. announced that a cranium from China of an archaic member of the genus *Homo*, said to have been hidden from the Japanese during World War II and known as "Dragon Man," was named the holotype of a new species, *Homo longi*. Although no DNA has yet been extracted from the 148,000-year-old hominin, some researchers suspect it could be a Denisovan, although its discoverers think it is more closely related to *H. sapiens* than either the Neandertals or the Denisovans are! Fahu Chen et al., "A Late Middle Pleistocene Denisovan Mandible from the Tibetan Plateau," *Nature* 569 (2019): 409–12; Qiang Ji et al., "Late Middle Pleistocene Harbin Cranium Represents New *Homo* Species," *Innovation* 2 (2021): 100132.

18. Kay Prüfer et al., "The Complete Genome Sequence of a Neanderthal from the Altai Mountains," *Nature* 504 (2014): 43–49.

19. Viviane Slon et al., "The Genome of the Offspring of a Neanderthal Mother and a Denisovan Father," *Nature* 561 (2018): 7721.

20. Qiaomei Fu et al., "Genome Sequence of a 45,000-Year-Old Modern Human from Western Siberia," *Nature* 514 (2014): 445–49.

21. Qiaomei Fu et al., "An Early Modern Human from Romania with a Recent Neanderthal Ancestor," *Nature* 524 (2015): 216–19.

22. Qiaomei Fu et al., "The Genetic History of Ice Age Europe," *Nature* 534 (2016): 200–205.

23. Mateja Hajdinjak et al., "Initial Upper Palaeolithic Humans in Europe Had Recent Neanderthal Ancestry," *Nature* 592 (2021): 253–57.

24. Kay Prüfer et al., "A Genome Sequence from a Modern Human Skull over 45,000 Years Old from Zlatý kůň in Czechia," *Nature Ecology & Evolution* 5 (2021): 820–82.

25. Green et al., "A Draft Sequence of the Neandertal Genome," 721.

7. IS THERE SUCH A THING AS MODERN HUMAN BEHAVIOR?

1. Filé is ground dried sassafras leaves used as a thickening agent for soups by indigenous groups of the southeast United States; it was enthusiastically adopted by the French-speaking people of Louisiana.

2. Although behavior is used in descriptions of extant animal species, I urge caution when applying a behavioral "template" to *H. sapiens*. Evolution is defined as changes in the genetic properties of populations *across* generations. The wonderful thing about behavior is that it is so malleable it can change within the lifetime of an individual. This behavioral plasticity is particularly true of humans; we are the ultimate generalists, veritable "Jacks and Jills of all trades." Humans are amazing inventors and resource extractors whose cultures and technologies can change with dizzying speed. Although the capacity for our behavioral flexibility was almost certainly fashioned by selection, the archaeological evidence for this capacity is manifest in myriad ways.

3. April Nowell, "Defining Behavioral Modernity in the Context of Neandertal and Anatomically Modern Human Populations," *Annual Review of Anthropology* 39 (2010): 437–52.

4. Randall White, "Rethinking the Middle/Upper Paleolithic Transition," *Current Anthropology* 23 (1982): 169–92.

5. Richard G. Klein, "Archeology and the Evolution of Human Behavior," *Evolutionary Anthropology* 9 (2000): 17–36.

6. Sally McBrearty and Allison S. Brooks, "The Revolution That Wasn't: A New Interpretation of the Origin of Modern Human Behavior," *Journal of Human Evolution* 39 (2000): 453–563.

7. Pierre-Jean Texier et al., "A Howiesons Poort Tradition of Engraving Ostrich Eggshell Containers Dated to 60,000 Years Ago at Diepkloof Rock Shelter, South Africa," *Proceedings of the National Academy of Sciences* (USA) 107 (2010): 6180–85.

8. Sally McBrearty, "Comment to Henshilwood and Marean," *Current Anthropology* 44 (2003): 642.

9. Francesco D'Errico, "The Invisible Frontier: A Multiple Species Model for the Origin of Behavioral Modernity," *Evolutionary Anthropology* 12 (2003): 188–202.

10. Iain Davidson, "The Colonization of Australia and Its Adjacent Islands and the Evolution of Modern Cognition," *Current Anthropology* 51 (2010; Suppl.): S177–89.

11. The very late date (about 236–335 ka BP) by Dirks et al. for the recently discovered species *Homo naledi* by Berger et al. in South Africa (on which I had the privilege to work), emphasizes this point. *Homo naledi* is an archaic species with a brain less than half the size of human brains today, but it is present in Africa as recently as 236,000 years ago! When one finds a 250,000-year-old archaeological site in South Africa, how does one know it is not the work of this species? Paul Dirks et al., "The Age of *Homo naledi* and Associated Sediments in the Rising Star Cave, South Africa," *eLife* 6 (2017): e24231; Lee Berger et al., "*Homo naledi*, a New Species of the Genus *Homo* from the Dinaledi Chamber, South Africa," *eLife* 4 (2015): e09560.

12. Christopher S. Henshilwood and Curtis W. Marean, "The Origin of Modern Human Behavior: Critique of the Models and Their Test Implications," *Current Anthropology* 44 (2003): 627–51.

13. Trenton W. Holliday, "Comment to Henshilwood and Marean," *Current Anthropology* 44 (2003): 639–40.

14. Davorka Radovčić et al., "Surface Analysis of an Eagle Talon from Krapina," *Nature Scientific Reports* 10 (2020): 6329.

15. Eugène Morin and Véronique Laroulandie, "Presumed Symbolic Use of Diurnal Raptors by Neanderthals," *PLoS ONE* 7 (2012): e32856.

16. Marco Peresani et al., "Late Neandertals and the Intentional Removal of Feathers as Evidenced from Bird Bone Taphonomy at Fumane Cave 44 ky B.P., Italy," *Proceedings of the National Academy of Sciences (USA)* 108 (2011): 3888–93; Morin and Laroulandie, "Presumed Symbolic Use of Diurnal Raptors by Neanderthals."

17. D'Errico, "The Invisible Frontier."

18. João Zilhão et al., "Symbolic Use of Marine Shells and Mineral Pigments by Iberian Neandertals," *Proceedings of the National Academy of Sciences (USA)* 107 (2010): 1023–28.

19. Dirk Hoffmann et al., "Symbolic Use of Marine Shells and Mineral Pigments by Iberian Neandertals 115,000 Years Ago," *Science Advances* 4, no. 2 (2018): eaar5255.

20. Dirk Hoffmann et al., "U-Th Dating of Carbonate Crusts Reveals Neandertal Origin of Iberian Cave Art," *Science* 359 (2018): 912–15; Dirk Leder et al., "A 51,000-Year-Old Engraved Bone Reveals Neanderthals' Capacity for Symbolic Behaviour," *Nature Ecology & Evolution* 5 (2021): 1273–82.

21. Jacques Jaubert et al., "Early Neanderthal Constructions Deep in Bruniquel Cave in Southwestern France," *Nature* 534 (2016): 111–14.

22. D'Errico, "The Invisible Frontier."

23. João Zilhão, "Comment to Henshilwood and Marean," *Current Anthropology* 44 (2003): 643.

24. Zilhão, "Comment to Henshilwood and Marean."

25. John J. Shea, "*Homo sapiens* Is as *Homo sapiens* Was: Behavioral Variability Versus 'Behavioral Modernity' in Paleolithic Archaeology," *Current Anthropology* 52 (2011): 1–35.

26. Statistically speaking, a mode is a peak in a Gaussian distribution (bell curve). It's the value with the most observations in a sample. In a normal distribution, the mode is coincident with both the mean, or arithmetic average, and the median, which is the observation falling at the middle of the distribution (50th percentile). Not all distributions are normal. Skew refers to a nonnormal distribution in

which there is a long tail to one side of the mean. Finally, the variance of a distribution is a measure of dispersion about the mean—a flatter, broader bell curve has greater variance than a taller, narrower bell curve.

27. Shea, "*Homo sapiens* Is as *Homo sapiens* Was," 7.

28. Grahame Clark, *World Prehistory: A New Outline* (Cambridge: Cambridge University Press, 1969).

29. Metin I. Eren, "Comment to Shea," *Current Anthropology* 52 (2011): 18; April Nowell, "Comment to Shea," *Current Anthropology* 52 (2011): 19–20.

30. You intuitively know this to be true. For example, if I were to tell you that the distance from New Orleans to the Louisiana city of Hammond is 58 miles, plus or minus a quarter of a mile, you wouldn't bat an eyelash. In contrast, if I were to tell you that I am 6′ 1″ tall, plus or minus a quarter of a mile, you would think something was wrong with me.

31. To maintain 1.0 as the isometric slope when comparing a linear measurement to body mass (i.e., a single dimensional measure compared to a three-dimensional one), one uses the cube root of body mass as the independent variable. For the allometric relationship between an area measurement to body mass (i.e., a two-dimensional measure to a three-dimensional one), body mass to the 2/3 power is the independent variable of choice.

32. A. M. Boddy et al., "Comparative Analysis of Encephalization in Mammals Reveals Relaxed Constraints on Anthropoid Primate and Cetacean Brain Scaling," *Journal of Evolutionary Biology* 25 (2012): 981–94.

33. Boddy et al., "Comparative Analysis of Encephalization in Mammals."

34. Este Armstrong, "Brains, Bodies and Metabolism," *Brain, Behavior and Evolution* 36 (1990): 166–76.

35. Harry J. Jerison, *Evolution of the Brain and Intelligence* (New York: Academic Press, 1973).

36. Boddy et al., "Comparative Analysis of Encephalization in Mammals."

37. Given the example of baleen whales, in which their growth to an extremely large size in the absence of brain growth leads to a much smaller EQ, perhaps a similar phenomenon (albeit not as extreme) is at work here too because gorillas are by far the largest extant primate species.

38. Christopher B. Ruff, Erik Trinkaus, and Trenton W. Holliday, "Body Mass and Encephalization in Pleistocene *Homo*," *Nature* 387 (1997): 173–76.

39. The femoral head supports the weight of the trunk, head, neck, and upper limbs during locomotion. As such, it is highly correlated with body mass. However, when people are more active as subadults, their femoral heads grow larger. This is because during strenuous activities the lower limb experiences great biomechanical stresses—forces up to five times one's body weight—which stimulate bone growth (chapter 9). Femoral head size could therefore make active people look heavier than less active people. Given that fossil hominins were probably more active than most people today, using bi-iliac breadth and stature to estimate body mass is an independent, nonbiomechanical check on this potential bias.

40. Robert D. Martin, "Relative Brain Size and Basal Metabolic Rate in Terrestrial Vertebrates," *Nature* 293 (1981): 57–60.

41. Boddy et al., "Comparative Analysis of Encephalization in Mammals."

42. Boddy et al., "Comparative Analysis of Encephalization in Mammals."

43. Boddy et al., "Comparative Analysis of Encephalization in Mammals."

44. Armstrong, "Brains, Bodies and Metabolism."

45. The type specimen of *H. floresiensis*, Liang Bua 1 (LB1), dating to about 60,000 to 100,000 years ago, is of Late Pleistocene age. However, given its small-brained morphology and plesiomorphic postcranial skeleton, its ancestors probably split from ours in the Early Pleistocene. I have therefore included it among the Middle Pleistocene hominins.

46. Gerhard Roth and Ursula Dicke, "Evolution of the Brain and Intelligence," *Trends in Cognitive Sciences* 9 (2005): 250–57.

47.   The cerebral cortex is the outermost portion of the cerebrum, itself the largest part of the human brain. The cortex is gray matter, primarily comprised of cell bodies of neurons and their support-ing glial cells. The cortex overlies white matter, which is comprised of axons of neurons that are myelinated (covered with a whitish substance that aids in the rapid transmission of electrical signals between cells). Deeper still within the white matter sit pockets of gray matter known as basal nuclei. From fossil skulls, it is impossible to determine the thickness of the cerebral cortex, its size vs. size of the white matter, or basal nuclei.

48.   Cleber A. Trujillo et al., "Reintroduction of the Archaic Variant of NOVA1 in Cortical Organoids Alters Neurodevelopment," *Science* 371 (2021): eaax2537.

49.   Ezra Zubrow, "The Demographic Modeling of Neanderthal Extinction," in *The Human Revolution: Behavioural, and Biological Perspectives on the Origins of Modern Humans*, ed. Paul Mellars and Chris Stringer (Edinburgh: University of Edinburgh Press, 1989), 212–31.

50.   Steven E. Churchill, *Thin on the Ground: Neandertal Biology, Archeology, and Ecology* (Hoboken, NJ: Wiley Blackwell, 2014).

51.   Higher population densities make the development and spread of technological innovations much more rapid. In this light, the fact that modern humans were able to colonize areas (Siberia north of the Arctic Circle, for example) in which the Neandertals (and Denisovans) had been unable to thrive could simply be due to the fact that modern humans show demographic expansion beyond Africa already possessing more extensive social networks that promote rapid cultural and technological exchange.

8. NEANDERTAL AND CRO-MAGNON INTERACTIONS IN EUROPE

1.   Ludovic Slimak et al., "Modern Human Incursion into Neanderthal Territories 54,000 Years Ago at Mandrin, France," *Science Advances* 8 (2022): eabj9496.

2.   Helen Fewlass et al., "A ^{14}C Chronology for the Middle to Upper Palaeolithic Transition at Bacho Kiro Cave, Bulgaria," *Nature Ecology & Evolution* 4 (2020): 794–801.

3.   Zooarchaeology is the study of animal remains recovered from archaeological contexts. Specialists are trained in much the same way as vertebrate or invertebrate paleontologists, at least in terms of identifying species in the fossil record from fragmentary remains. The difference is that zooarchae-ologists are more interested in what humans are doing with the animals in question, or what the animals can tell us about the people of a particular site, than addressing questions about the animals themselves.

4.   Jean-Jacques Hublin et al., "Initial Upper Palaeolithic *Homo sapiens* from Bacho Kiro Cave, Bul-garia," *Nature* 581 (2020): 299–302.

5.   Hublin et al., "Initial Upper Palaeolithic *Homo sapiens* from Bacho Kiro Cave, Bulgaria."

6.   João Zilhão argued that the late Neandertals from Level G1 at the Croatian site of Vindija are asso-ciated with the Szeletian industry—this to counter a claim by Ivor Karavanić that the Vindija Nean-dertals are associated with a regional variant of the Aurignacian. Zilhão reports direct (albeit not ultrafiltrated) radiocarbon dates on two Neandertal specimens from Vindija ranging from 32,540 ± 370 to 37,360 ± 2240 cal BP. João Zilhão, "Szeletian, Not Aurignacian: A Review of the Chronol-ogy and Cultural Associations of the Vindija G1 Neandertals," in *Sourcebook of Paleolithic Transi-tions*, ed. Marta Camps and Parth Chauhan (New York: Springer, 2009), 407–26; Ivor Karavanić, "Olschewian and Appearance of Bone Technology in Croatia and Slovenia," in *Neanderthals and Modern Humans—Discussing the Transition. Central and Eastern Europe from 50.000—30.000 B.P.*, ed. Jorg Örschiedt and Gerd-Christian Weniger (Mettmann, Neandertal Museum, 2000), 159–68.

7. Gilbert Tostevin puts forward the provocative hypothesis that the Bohunician is an intrusive indus-
    try made by modern humans who, by coming into contact with Neandertals making Micoquian
    Middle Paleolithic tools, inspired Neandertals to invent the Szeletian. Gilbert Tostevin, "Social
    Intimacy, Artefact Visibility, and Acculturation Models of Neanderthal-Modern Human Interac-
    tion," in *Rethinking the Human Revolution: New Behavioural and Biological Perspectives on the Origin
    and Dispersal of Modern Humans*, ed. Paul Mellars, Katie Boyle, Ofer Bar-Yosef, and Chris Stringer
    (Cambridge: MacDonald Institute Monographs, 2007), 341–57.

8. Paola Villa et al., "From Neandertals to Modern Humans: New Data on the Uluzzian," *PLoS ONE*
    13 (2018): e0196786.

9. Francesco D'Errico, et al., "The Invisible Frontier: A Multiple Species Model for the Origin of
    Behavioral Modernity," *Evolutionary Anthropology* 12 (2003): 188–202.

10. Stefano Benazzi et al., "Early Dispersal of Modern Humans in Europe and Implications for Neander-
    thal Behaviour," *Nature* 479 (2011): 525–28; João Zilhão, William E. Banks, Francesco d'Errico, and
    Patrizia Gioia, "Analysis of Site Formation and Assemblage Integrity Does Not Support Attribution
    of the Uluzzian to Modern Humans at Grotta del Cavallo," *PLoS ONE* 10 (2015): e0131181.

11. Even if the levels at Cavallo are mixed, we still have the awl from Uluzzian levels at La Fabbrica.

12. Roxane Rocca, Nelly Connet, and Vincent Lhomme, "Before the Transition? The Final Middle Palae-
    olithic Lithic Industry from the Grotte du Renne (LayerXI) at Arcy-sur-Cure (Burgundy, France),"
    Comptes Rendus Palevol 16 (2017): 878–93. In the book *Kindred*, the archaeologist Rebecca Wragg
    Sykes argues that only in older excavations do Châtelperronian levels have a significant Middle Paleo-
    lithic component. She says that in recent excavations the Middle Paleolithic element is absent, imply-
    ing older excavators were not recognizing the mixing of Mousterian and Châtelperronian levels. Her
    book lacks in-text references, so I am uncertain of the excavations to which she is referring. Rebecca
    Wragg Sykes, *Kindred: Neanderthal Life, Love, Death and Art* (New York: Bloomsbury Sigma, 2020).

13. Jean-Jacques Hublin et al., "A Late Neanderthal Associated with Upper Paleolithic Artefacts,"
    Nature 381 (1996): 224–26.

14. Frido Welker et al., "Palaeoproteomic Evidence Identifies Archaic Hominins Associated with the
    Châtelperronian at the Grotte du Renne," *Proceedings of the National Academy of Sciences (USA)* 113
    (2016) 11162–67.

15. Francine David et al., "Le Châtelperronien de la Grotte du Renne à Arcy-sur-Cure (Yonne). Don-
    nées sédimentologiques et chronostratigraphiques," *Bulletin de la Société Préhistorique Française* 98
    (2001): 207–30; Ofer Bar-Yosef and Jean-Guillaume Bordes, "Who Were the Makers of the Châtel-
    perronian Culture?," *Journal of Human Evolution* 59 (2010): 586–93.

16. Bar-Yosef and Bordes, "Who Were the Makers of the Châtelperronian Culture?"

17. Bar-Yosef and Bordes, "Who Were the Makers of the Châtelperronian Culture?"; Brad Gravina
    et al., "No Reliable Evidence for a Neanderthal-Châtelperronian Association at La Roche-à-Pierrot,
    Saint-Césaire," *Nature Scientific Reports* 8 (2018): 15134.

18. François Caron et al., "The Reality of Neandertal Symbolic Behavior at the Grotte du Renne,
    Arcy-sur-Cure, France," *PLoS ONE* 6 (2011): e21545.

19. Jean-Jacques Hublin et al., "Radiocarbon Dates from the Grotte du Renne and Saint-Césaire Sup-
    port a Neandertal Origin for the Châtelperronian," *Proceedings of the National Academy of Sciences
    (USA)* 109 (2012): 18743–48.

20. James L. Bischoff, Narcís Soler, and Julià Maroto Ramon, "Abrupt Mousterian/Aurignacian Cound-
    ary at c. 40 ka BP: Accelerator 14C dates from l'Arbreda Cave (Catalunya, Spain)," *Journal of Archae-
    ological Science* 16 (1989): 563–76; Victoria Valdes Cabrera and James L. Bischoff, "Accelerator ^{14}C
    Dates for Early Upper Paleolithic (Basal Aurignacian) at El Castillo Cave (Spain)," *Journal of Archae-
    ological Science* 16 (1989): 577–84.

21. R. E. Wood et al., "The Chronology of the Earliest Upper Palaeolithic in Northern Iberia: New
    Insights from L'Arbreda, Labeko Koba and La Viña," *Journal of Human Evolution* 69 (2014):

91–109; Rachel Wood et al., "El Castillo (Cantabria, Northern Iberia) and the Transitional Aurignacian: Using Radiocarbon Dating to Assess Site Taphonomy," *Quaternary International* 474 (2018): 56–70.

El Castillo is the only site at which an industry called the Transitional Aurignacian was found. An archaeological level attributed to this industry (Unit 18) was dated by Wood et al. in 2018. The problem is that some researchers (compare Zilhão and D'Errico) have argued that the Transitional Aurignacian is not a real entity but rather a mix of Mousterian and Aurignacian levels. João Zilhão and Francesco d'Errico, "An Aurignacian Garden of Eden in Southern Germany? An Alternative Interpretation of the Geissenklösterle and a Critique of the Kulturpumpe Model," *Paleo* 15 (2003): 1–28.

22. Philip R. Nigst et al., "Early Modern Human Settlement of Europe North of the Alps Occurred 43,500 Years Ago in a Cold Steppe-Type Environment," *Proceedings of the National Academy of Sciences (USA)* 111 (2014): 14394–99.

23. Thomas Higham et al., "Testing Models for the Beginnings of the Aurignacian and the Advent of Figurative Art and Music: The Radiocarbon Chronology of Geißenklösterle," *Journal of Human Evolution* 62 (2012): 664–76.

24. The sample taken from the Feldhofer Neandertal humerus for radiocarbon dating was removed from the internal surface of the bone and therefore should be less contaminated than its external surface. However, it is not ultrafiltrated, so it is possible this date is too young. Ralf W. Schmitz et al., "The Neandertal Type Site Revisited: Interdisciplinary Investigations of Skeletal Remains from the Neander Valley, Germany," *Proceedings of the National Academy of Sciences (USA)* 99 (2002): 13342–47.

25. Armando Falcucci, Nicholas J. Conard, and Marco Peresani, "A Critical Assessment of the Protoaurignacian Lithic Technology at Fumane Cave and Its Implications for the Definition of the Earliest Aurignacian," *PLoS ONE* 12 (2017): e0189241.

26. Katerina Douka et al., "A New Chronostratigraphic Framework for the Upper Palaeolithic of Riparo Mochi (Italy)," *Journal of Human Evolution* 62 (2012): 286–99.

27. Lars Anderson, Natasha Reynolds, and Nicolas Teyssandier, "No Reliable Evidence for a Very Early Aurignacian in Southern Iberia," *Nature Ecology & Evolution* 3 (2019): 713.

28. Jonathan A. Haws et al., "The Early Aurignacian Dispersal of Modern Humans into Westernmost Eurasia," *Proceedings of the National Academy of Sciences (USA)* 117 (2020): 25414–22.

29. There are many "Grottes des Fées" (Fairy Caves) in France, some preserving archaeological levels. Here I am referring to the one in the commune of Châtelperron, the type site of the Châtelperronian.

30. Specifically, at El Pendo, a Châtelperronian layer was said to overlie two Aurignacian layers. At Grotte des Fées, two Aurignacian levels were said to overlie two Châtelperronian levels, and the Aurignacian levels were said to lie beneath four Châtelperronian levels. At Le Piage, a Châtelperronian layer was said to overlie three Aurignacian layers and lie beneath a fourth. At Roc de Combe, an Aurignacian layer was said to be sandwiched between two Châtelperronian levels; the top Châtelperronian level was then itself overlain by another Aurignacian layer.

31. I suppose finding a burial pit with a Neandertal and a Cro-Magnon side-by-side in a loving embrace, the former holding a Châtelperron knife and the latter a split-based bone point (or vice versa), might be too much to ask.

32. Jean-Guillaume Bordes, "Les interstratifications Châtelperronien / Aurignacien du Roc-de-Combe et du Piage (Lot, France). Analyse taphonomique des industries lithiques; implications archéologiques" (PhD diss. Université de Bordeaux I, 2002); Ramón Montes et al., "La secuencia estratigráfica de la cueva de El Pendo (Escobedo de Camargo, Cantabria): Problemas geoarqueológicos de un referente cronocultural," in *Geoarqueología y patrimonio en la Península Ibérica y el entorno mediterráneo*, ed. Manuel Santonja, Alfredo Pérez-González, and María José Machado (Almazán: ADEMA, 2005), 139–59.

33. João Zilhão et al., "Grotte des Fées (Châtelperron): History of Research, Stratigraphy, Dating, and Archaeology of the Châtelperronian Type-Site," *PaleoAnthropology* 2008: 1–42.

34. Paul Mellars and Brad Gravina, "Châtelperron: Theoretical Agendas, Archaeological Facts, and Diversionary Smoke-Screens," *PaleoAnthropology* 2008: 43–64.

35. Julien Riel-Salvatore, Alexandra E. Miller, and Geoffrey A. Clark, "An Empirical Evaluation of the Case for a Châtelperronian-Aurignacian Interstratification at Grotte des Fées de Châtelperron," *World Archaeology* 40 (2008): 480–92.

36. Paul Mellars, "The Impossible Coincidence. A Single-Species Model for the Origins of Modern Human Behavior in Europe," *Evolutionary Anthropology* 14 (2005): 12–27.

37. Francesco D'Errico et al., "Neanderthal Acculturation in Western Europe? A Critical Review of the Evidence and Its Interpretation," *Current Anthropology* 39 (1998; Suppl.): S1–S44; João Zilhão et al., "Analysis of Aurignacian Interstratification at the Châtelperronian-Type Site and Implications for the Behavioral Modernity of Neandertals," *Proceedings of the National Academy of Sciences (USA)* 103 (2006): 12643–48.

38. Compare Zilhão and d'Errico, "An Aurignacian Garden of Eden in Southern Germany?"

39. Using tool types to assess the biological affinities of their makers is treacherous territory indeed. I am reminded of Dan Lieberman's quip that the fact he has a wok in his kitchen does not establish Chinese ancestry for him.

40. Alfred L. Kroeber, "Sub-Human Cultural Beginnings," *Quarterly Review of Biology* 3 (1928): 325–42.

41. I would argue that the perforated shells from 50,000-year-old Mousterian contexts at the two Spanish sites of Cueva de los Aviones and Cueva Antón (chapter 7) prove Neandertals independently came up with the idea of piercing shells for ornamentation. However, given that we now have 54,000-year-old modern humans at Mandrin Cave in France, modern humans could very well have been in Spain about 50,000 years ago. If the recent 115 ka BP U-series date at Aviones stands up to further scrutiny, however, then to my mind the debate is over.

42. Clive Finlayson, *Neanderthals and Modern Humans: An Ecological and Evolutionary Perspective* (Cambridge: Cambridge University Press, 2004).

43. Compare D'Errico et al., "Neanderthal Acculturation in Western Europe?"

44. Finlayson, *Neanderthals and Modern Humans.*

45. Jean-Jacques Hublin et al., "The Mousterian Site of Zafarraya (Andalucia, Spain): Dating and Implications on the Palaeolithic Peopling Processes of Western Europe." *Comptes Rendus de l'Academie de Sciences II* 32 (1995): 931–37.

46. Santiago Fernández et al., "The Holocene and Upper Pleistocene Pollen Sequence of Carihuela Cave, Southern Spain," *Geobios* 40 (2007): 75–90.

47. João Zilhão and Paul Pettitt, "On the New Dates for Gorham's Cave and the Late Survival of Iberian Neanderthals," *Before Farming* 3 (2006): 1–9.

48. João Zilhão, "Chronostratigraphy of the Middle-to-Upper Paleolithic Transition in the Iberian Peninsula," *Pyrenae* 37 (2006): 7–84.

49. Diego E. Angelucci and João Zilhão, "Stratigraphy and Formation Processes of the Upper Pleistocene Deposit at Gruta da Oliveira, Almonda Karstic System, Torres Novas, Portugal," *Geoarchaeology: An International Journal* 24 (2009): 277–310.

50. Martin Kehl et al., "Late Neanderthals at Jarama VI (Central Iberia)?," *Quaternary Research* 80 (2013): 218–34.

51. Rachel E. Wood et al., "Radiocarbon Dating Casts Doubt on the Late Chronology of the Middle to Upper Palaeolithic Transition in Southern Iberia," *Proceedings of the National Academy of Sciences (USA)* 110 (2013): 2781–86.

52. Haws et al., "The Early Aurignacian Dispersal of Modern Humans into Westernmost Eurasia."

53. There is another possibility here. Recall that I worked at two Gravettian-aged sites in Portugal for which the lithic remains were decidedly Middle Paleolithic in character. The dates for the "late Mousterian" in Gorham's Cave could therefore be accurate, but the materials were left behind by Gravettian-wielding modern humans.

9. BIOANTHROPOLOGY OF THE CRO-MAGNONS

1. Georges Cuvier, *Le Règne Animal* (Paris: A. Belin, Chez Déterville, 1817).

2. Johann Friedrich Blumenbach, *De Generis Humani Varietate Nativa*, 3rd ed. (Göttingen: Vandenhoek and Ruprecht, 1795).

3. Earnest Albert Hooton, "Methods of Racial Analysis," *Science*, n.s., 63 (1926): 75–81, at 76. Whether we like it or not, most PhD biological anthropologists in the United States trace our academic roots to Earnest Hooton, who for many years was one of the few U.S. scholars supervising PhD students in the subject. My own academic roots (i.e., my PhD advisor back to his PhD advisor, etc., trace to Hooton as follows: Erik Trinkaus, Alan Mann, Sherwood Washburn, Earnest Hooton). As an aside, my four bioanthropology colleagues at Tulane (Katharine Jack, a primatologist; Katharine Lee, a biomedical anthropologist; Jennifer Spence, a human biologist and forensic anthropologist; and John Verano, a bioarchaeologist) all trace their academic roots to Hooton as well.

4. Essentialism is the Platonic notion that we recognize categories of things because everything in the universe has its own underlying, immutable, eternal, "essence." This essence cannot be directly observed, and all observable objects are imperfect replicas of this ideal.

5. Earnest Albert Hooton, *Up From the Ape* (New York: Macmillan, 1935).

6. Carleton S. Coon, *The Origin of Races* (New York: Alfred A. Knopf, 1962). If you are skeptical of this claim, I invite you to find a copy of Coon's book in your local library, turn to the photographic plates in the middle of the book, and look at the last one (Plate 32).

7. Stanley M. Garn, *Human Races* (Springfield, IL: Charles C. Thomas, 1961).

8. R. C. Lewontin, "The Apportionment of Human Diversity," in *Evolutionary Biology*, ed. Theodosius Dobzhansky, Max K. Hecht, and William C. Steere (New York: Springer, 1972), 381–98.

9. Jeffrey C. Long, Jie Li, and Meghan E. Healy, "Human DNA Sequences: More Variation and Less Race," *American Journal of Physical Anthropology* 139 (2009): 23–34.

10. Ernst Mayr, *Principles of Systematic Zoology* (New York: McGraw-Hill, 1969).

11. Anne Fischer, Victor Wiebe, Svante Pääbo, and Molly Przeworski, "Evidence for a Complex Demographic History of Chimpanzees," *Molecular Biology and Evolution* 21 (2004): 799–808.

12. René Verneau, *Les Grottes de Grimaldi (Baoussé-Roussé). Anthropologie* (Monaco: Imprimerie de Monaco, 1906). Verneau's Grimaldi Race was based on his own reconstruction errors as well as misinterpretation of the morphology of the Grotte des Enfants 5 and 6 skeletons.

13. Sara D. Niedbalski and Jeffrey C. Long, "Novel Alleles Gained During the Beringian Isolation Period," *Nature Scientific Reports* 12 (2022): 4289; Heather L. Norton et al., "Genetic Evidence for the Convergent Evolution of Light Skin in Europeans and East Asians," *Molecular Biology and Evolution* 24 (2007): 710–22.

14. Sandra Wilde et al., "Direct Evidence for Positive Selection of Skin, Hair, and Eye Pigmentation in Europeans During the Last 5,000 Years," *Proceedings of the National Academy of Sciences (USA)* 111 (2014): 4832–37.

15. Paul Broca, "Sur les crânes et ossements des Eyzies." *Bulletins de la Société d'Anthropologie de Paris* 3 (1868): 350–92.

16. Philippe Charlier et al., "Did Cro-Magnon 1 have Neurofibromatosis Type 1?" *The Lancet* 391 (2018): 1259. The flat bones of the cranial vault are organized in such a way that they have outer and inner "tables" of dense cortical bone, with a spongy bone interior called diploë sandwiched between them that is filled with red bone marrow. I used to tell students I was going to name my first daughter Diploë. I didn't.

17. Verneau, *Les Grottes de Grimaldi (Baoussé-Roussé).*

18. Erik Trinkaus, "Body Length and Mass," in *Early Modern Human Evolution in Central Europe: The People of Dolní Věstonice and Pavlov*, ed. Erik Trinkaus and Jiří Svoboda (Oxford: Oxford University

Press, 2006), 233–41; Emanuel Vlček, *Die Mammutjäger von Dolní Věstonice* (Liestal, Switzerland: Archäologie und Museum 22, 1991).

19. In Europe, bioarchaeology is a more general term, referring to the recovery and analysis of the remains of any zoological (including human) or botanical remains from archaeological contexts.

20. Simon W. Hillson, Robert G. Franciscus, Trenton W. Holliday, and Erik Trinkaus, "The Ages at Death," in *Early Modern Human Evolution in Central Europe: The People of Dolní Věstonice and Pavlov*, ed. Erik Trinkaus and Jiří Svoboda (Oxford: Oxford University Press, 2006), 31–45.

21. Alissa Mittnik, Chuan-Chao Wang, Jiří Svoboda, and Johannes Krause, "A Molecular Approach to the Sexing of the Triple Burial at the Upper Paleolithic Site of Dolní Věstonice," *PLoS ONE* 11 (2016): e0163019.

22. Erik Trinkaus, "An Abundance of Developmental Anomalies and Abnormalities in Pleistocene People," *Proceedings of the National Academy of Sciences (USA)* 115 (2018): 11941–46.

23. The shaft of a long bone is hollow; the hollow is called the medullary cavity, which in living adults is filled with fatty yellow marrow. Surrounding the medullary cavity is dense bony tissue known as cortical bone, or cortex.

24. Erik Trinkaus, Simon W. Hillson, Robert G. Franciscus, and Trenton W. Holliday, "Skeletal and Dental Paleopathology," in *Early Modern Human Evolution in Central Europe: The People of Dolní Věstonice and Pavlov*, ed. Erik Trinkaus and Jiří Svoboda (Oxford: Oxford University Press, 2006), 419–58.

25. Emanuel Vlček, *Die Mammutjäger von Dolní Věstonice*; Trinkaus et al., "Skeletal and Dental Paleopathology."

26. Steven E. Churchill and Vincenzo Formicola, "A Case of Marked Bilateral Asymmetry in the Upper Limbs of an Upper Palaeolithic Male from Barma Grande (Liguria), Italy," *International Journal of Osteoarchaeology* 7 (1997): 18–38. The cortical bone's cross-sectional area is measured from either biplanar radiographs or CT scans in a plane perpendicular to the long axis of the shaft. In this case, the cortical area is measured at the humerus's 35 percent section, which is located 35 percent of the way up the bone (about 3 to 4 inches) from its distal or elbow end.

27. Vitale S. Sparacello, Sébastien Villotte, Laura L. Shackelford, and Erik Trinkaus, "Patterns of Humeral Asymmetry Among Late Pleistocene Humans," *Comptes Rendus Palevol* 16 (2017): 680–89.

10. SLINGS AND ARROWS

1. Harmut Thieme, "Lower Palaeolithic Throwing Spears and Other Wooden Implements from Schöningen," in *Hominid Evolution: Lifestyles and Survival Strategies*, ed. Herbert Ullrich (Gelsenkirchen: Archaea, 1999), 383–95; Hartmut Thieme, "Lower Palaeolithic Hunting Spears from Germany," *Nature* 385 (1997): 807–10.

2. John J. Shea, "The Origins of Lithic Projectile Point Technology: Evidence from Africa, the Levant, and Europe," *Journal of Archaeological Science* 33 (2006): 823–46; Steven E. Churchill, *Thin on the Ground: Neandertal Biology, Archeology and Ecology* (Hoboken, NJ: Wiley Blackwell, 2014).

3. Churchill, *Thin on the Ground*, 63.

4. Steven E. Churchill, "Weapon Technology, Prey Size Selection, and Hunting Methods in Modern Hunter-Gatherers: Implications for Hunting in the Palaeolithic and Mesolithic," in *Hunting and Animal Exploitation in the Later Palaeolithic and Mesolithic of Eurasia*, ed. Gail L. Peterkin, Harvey Bricker, and Paul A. Mellars (Washington, DC: American Anthropological Association, 1993), 11–24.

5. Churchill, *Thin on the Ground*.

6. Shea, "The Origins of Lithic Projectile Point Technology." Darts are projectiles launched from spear throwers, frequently called atlatls in North America. I discuss these later in the chapter.

7. Susan S. Hughes, "Getting to the Point: Evolutionary Change in Prehistoric Weaponry," *Journal of Archaeological Method and Theory* 5 (1998): 345–408.

8. Marlize Lombard, "Quartz-Tipped Arrows Older Than 60 ka: Further Use-Trace Evidence from Sibudu, KwaZulu-Natal, South Africa," *Journal of Archaeological Science* 38 (2011): 1918–30.

9. Shea, "The Origins of Lithic Projectile Point Technology."

10. Katsuhiro Sano et al., "The Earliest Evidence for Mechanically Delivered Projectile Weapons in Europe," *Nature Ecology & Evolution* 3 (2019): 1409–14.

11. Elise Tartar and Randall White, "The Manufacture of Aurignacian Split-Based Points: An Experimental Challenge," *Journal of Archaeological Science* 40 (2013): 2723–45.

12. Churchill, "Weapon Technology, Prey Size Selection, and Hunting Methods in Modern Hunter-Gatherers."

13. Sano et al., "The Earliest Evidence for Mechanically Delivered Projectile Weapons in Europe."

14. Michelle Langley, "Late Pleistocene Osseous Projectile Technology and Cultural Variability," in *Osseous Projectile Weaponry: Towards an Understanding of Pleistocene Cultural Variability*, ed. Michelle C. Langley (Dordrecht: Springer, 2016), 1–11.

15. Churchill, "Weapon Technology, Prey Size Selection, and Hunting Methods in Modern Hunter-Gatherers."

16. Ethnoarchaeology is a field pioneered by Binford in which archaeologists live with a group of people as they engage in their traditional lifestyle. The object is to see how archaeological sites are formed by watching them being made in real time.

17. Lioudmila Iakovleva, "L'art mézinien en Europe orientale dans son context chronologique, culturel et spirituel," *L'Anthropologie* 113 (2009): 691–752.

18. Mietje Germonpré, Martina Lázničková-Galetová, and Mikhail V. Sablin, "Palaeolithic Dog Skulls at the Gravettian Předmostí Site, the Czech Republic," *Journal of Archaeological Science* 39 (2012): 184–202.

19. Pat Shipman, *The Invaders: How Humans and Their Dogs Drove Neanderthals to Extinction* (Cambridge, MA: Belknap Press of Harvard University Press, 2015).

20. Hervé Bocherens et al., "Reconstruction of the Gravettian Food-Web at Předmostí I Using Multi-Isotopic Tracking (^{13}C, ^{15}N, ^{34}S) of Bone Collagen," *Quaternary International* 359–360 (2015): 211–28.

21. Michael P. Richards, Paul B. Pettitt, Mary C. Stiner, and Erik Trinkaus, "Stable Isotope Evidence for Increasing Dietary Breadth in the European Mid-Upper Paleolithic," *Proceedings of the National Academy of Sciences (USA)* 98 (2001): 6528–32.

22. Hervé Bocherens, Gennady Baryshnikov, and Wim Van Neer, "Were Bears or Lions Involved in Salmon Accumulation in the Middle Palaeolithic of the Caucasus? An Isotopic Investigation in Kudaro 3 Cave," *Quaternary International* 339–340 (2014): 112–18.

23. Christoph Wissing et al., "Stable Isotopes Reveal Patterns of Diet and Mobility in the Last Neandertals and First Modern Humans in Europe," *Nature Scientific Reports* 9 (2019): 4433.

24. Bocherens et al., "Reconstruction of the Gravettian Food-Web at Předmostí I."

25. Rebeca García-González et al., "Dietary Inferences Through Dental Microwear and Isotope Analyses of the Lower Magdalenian Individual from El Mirón Cave (Cantabria, Spain)," *Journal of Archaeological Science* 60: (2015): 28–38.

26. Pierre-Jean Dodat et al., "Isotopic Calcium Biogeochemistry of MIS 5 Fossil Vertebrate Bones: Application to the Study of the Dietary Reconstruction of Regourdou 1 Neandertal Fossil," *Journal of Human Evolution* 151 (2021): 102925.

27. Karen Hardy, Stephen Buckley, and Les Copeland, "Pleistocene Dental Calculus: Recovering Information on Paleolithic Food Items, Medicines, Paleoenvironment and Microbes," *Evolutionary Anthropology* 27 (2018): 234–46.

28. Emanuela Cristiani, Anita Radini, Marija Edinborough, and Dušan Borić, "Dental Calculus Reveals Mesolithic Foragers in the Balkans Consumed Domesticated Plant Foods," *Proceedings of the National Academy of Sciences (USA)* 113 (2016): 10298–303.

29. Anna Revedin et al., "New Technologies for Plant Food Processing in the Gravettian," *Quaternary International* 359–360 (2015): 77–88.

11. CRO-MAGNON ART

1. A purported Mousterian (Neandertal) flute from Divje Babe, Slovenia, cannot be distinguished from bones with puncture marks from carnivore gnawing. That said, it is possible that the first flutes were accidentally discovered by blowing into such punctured bones. Nicholas J. Conard, Maria Malina, and Susanne C. Münzel, "New Flutes Document the Earliest Musical Tradition in Southwestern Germany," *Nature* 460 (2009): 737–40.

2. Zdeňka Nerudová et al., "The Woman from the Dolní Věstonice 3 Burial: A New View of the Face Using Modern Technologies," *Archaeological and Anthropological Sciences* (2019) 11: 2527–38.

3. A. W. G. Pike et al., "U-Series Dating of Paleolithic Art in 11 Caves in Spain," *Science* 336 (2012): 1409–13.

4. Romain Mensan et al., "Une nouvelle découverte d'art parietal aurignacien *in situ* à l'abri Castanet (Dordogne, France): Context et datation," *Paleo* 23 (2012): 171–88.

5. Anita Quiles et al., "A High Precision Chronological Model for the Decorated Upper Paleolithic Cave of Chauvet-Pont d'Arc, Ardèche, France," *Proceedings of the National Academy of Sciences (USA)* 113 (2016): 4670–75.

6. Paul Pettitt and Paul Bahn, "An Alternative Chronology for the Art of Chauvet Cave," *Antiquity* 89 (2015): 542–53.

7. These same techniques are in evidence in the (presumably) much earlier art at Chauvet Cave.

8. According to my mother, this was a common occurrence when I was a preschooler. On a "Charlie Brown" day, I would refuse to answer to my own name: I would only respond if my parents called me Charlie Brown, and I called my parents "Lucy" and "Linus." On other days I was "Ducky Lucky," and the same rules applied: my parents were "Goosey Loosey" and "Gander Lander." This goes to show that there is a very fine line between cute and obnoxious.

9. April Nowell, *Growing Up in the Ice Age: Fossil and Archaeological Evidence of the Lived Lives of Plio-Pleistocene Children* (Oxford: Oxbow Books, 2021).

10. E. B. White and Katharine S. White, *A Subtreasury of American Humor* (New York: Coward-McCann, 1941).

12. COLD COMFORT

1. Jessica Tierney et al., "Glacial Cooling and Climate Sensitivity Revisited," *Nature* 584 (2020): 569–73.

2. Miikka Tallavaara et al., "Human Population Dynamics in Europe over the Last Glacial Maximum," *Proceedings of the National Academy of Sciences (USA)* 112 (2015): 8232–37.

3. This interstadial is also known as the Bølling interstadial, named for a Danish lake where it was first identified.

4. For example, Amara et al. say that the above ground plant biomass (AGB) in East African grasslands is about 2,000 kg/ha, which Keesing says supports an estimated ungulate biomass of about

44 kg/ha. In contrast, the Conservation Biology Institute suggests that the AGB for forests in the eastern United States is about 150,000 kg/ha, and Jordan et al. found that moose on Isle Royale show a biomass of about 6.76 kg/ha. Edward Amara et al., "Aboveground Biomass Distribution in a Multi-Use Savannah Landscape in Southeastern Kenya: Impact of Land Use and Fences," *Land* 9 (2020): 381; Conservation Biology Institute, "Data Basin," accessed August 5, 2021, databasin.org; Felicia Keesing, "Cryptic Consumers and the Ecology of an African Savanna," *BioScience* 50 (2000): 205–15; Peter A. Jordan, Daniel B. Botkin, and Michael L. Wolfe, "Biomass Dynamics in a Moose Population," *Ecology* 52 (1971): 147–52.

5. The geneticist Emilia Huerta-Sánchez and her colleagues demonstrated that this high-altitude adaptation found in modern-day Tibetans was inherited from a Denisovan ancestor. Emilia Huerta-Sánchez et al., "Altitude Adaptation in Tibetans Caused by Introgression of Denisovan-like DNA," *Nature* 512 (2014): 194–97.

6. Those trained in rhetoric may accuse me of the fallacy of making an argument from personal incredulity, but I believe it is justified here.

7. Sophie Jørgensen-Rideout, "Hot Rocks, Cold Data: An Examination of Fire Use at the Upper Palaeolithic Site of Abri Pataud" (RMA thesis, University of Leiden, 2019).

8. Rebecca Wragg Sykes, *Kindred: Neanderthal Life, Love, Death and Art* (New York: Bloomsbury Sigma, 2020).

9. Wragg Sykes, *Kindred*.

10. In the early years of this millennium, I was looking forward to analyses determining the age of the phylogenetic split of human head lice (*Pediculus humanus*) from human pubic lice (*Pthirus pubis*) because that might suggest the time at which our ancestors lost their long body hair, making the head and pubic areas separate "ecosystems." Unfortunately, human head lice are most closely related to lice that afflict chimpanzees, and human pubic lice are more closely related to those that afflict gorillas, making it difficult to use these species' divergences time to pinpoint when, exactly, the loss of human body hair occurred. It is, however, suggestive of intimate (but not necessarily sexual) relations among proto-humans and proto-gorillas—a fascinating topic in its own right.

11. It is perhaps no surprise that the first eyed needles come from Central Asia. Europe is surrounded on three sides by ocean. In Western Europe, one is never far from a coast, so it is blessed with an oceanic environment characterized by milder winters, cooler summers, and more gradual transitions between the seasons. However, as one moves east, the distance between the northern and southern seas tends to increase, and the climate becomes more continental, with cold winters, hot summers, and rapid transitions between the seasons. Central Asia is more continental still. It has harsh enough winters today; imagine how much harsher they would have been in the Pleistocene!

12. Trenton W. Holliday, "The Ecological Context of Trapping Among Recent Hunter-Gatherers: Implications for Subsistence in Terminal Pleistocene Europe," *Current Anthropology* 39 (1998): 711–20.

13. Erik Trinkaus, "Anatomical Evidence for the Antiquity of Human Footwear Use," *Journal of Archaeological Science* 32 (2005): 1515–26.

14. The cold phases known as Dryas Stadials are named for a genus of tundra flowers that became more numerous during these periods.

15. The Allerød Interstadial is named for a village in Denmark just outside of Copenhagen, where its existence was first documented.

16. Even animals that eat wood (termites being the most famous example) are only capable of digesting it via a symbiotic relationship with bacteria living in their guts.

17. We know that at least some Epipaleolithic people are descendants of the post-LGM Cro-Magnons because they belong to the same genetic cluster, the Villabruna cluster (chapter 6).

18. Lawrence Guy Straus, Geoffrey A. Clark, Jesus Altuna, and Jesus A. Ortea, "Ice-Age Subsistence in Northern Spain," *Scientific American* 242 (1980): 142–52.

Bibliography

Allen, Joel A. "The Influence of Physical Conditions in the Genesis of Species." *Radical Review* 1 (1877): 108–40.

Amara, Edward, Hari Adhikari, Janne Heiskanen, Mika Siljander, Martha Munyao, Patrick Omondi, and Petri Pellikka. "Aboveground Biomass Distribution in a Multi-Use Savannah Landscape in Southeastern Kenya: Impact of Land Use and Fences." *Land* 9 (2020): 381.

Anderson, Lars, Natasha Reynolds, and Nicolas Teyssandier. "No Reliable Evidence for a Very Early Aurignacian in Southern Iberia." *Nature Ecology & Evolution* 3 (2019): 713.

Angelucci, Diego E., and João Zilhão. "Stratigraphy and Formation Processes of the Upper Pleistocene Deposit at Gruta da Oliveira, Almonda Karstic System, Torres Novas, Portugal." *Geoarchaeology: An International Journal* 24 (2009): 277–310.

Armstrong, Este. "Brains, Bodies and Metabolism." *Brain, Behavior and Evolution* 36 (1990): 166–76.

Auel, Jean M. *The Clan of the Cave Bear*. New York: Bantam, 1980.

Bar-Yosef, Ofer, and Jean-Guillaume Bordes. "Who Were the Makers of the Châtelperronian Culture?" *Journal of Human Evolution* 59 (2010): 586–93.

Benazzi, Stefano, Katerina Douka, Cinzia Fornai, Catherine C. Bauer, Ottmar Kullmer, Jiří Svoboda, Ildikó Pap, et al. "Early Dispersal of Modern Humans in Europe and Implications for Neanderthal Behaviour." *Nature* 479 (2011): 525–28.

Berger, Lee R., John Hawks, Darryl J. de Ruiter, Steven E Churchill, Peter Schmid, Lucas K. Delezene, Tracy L. Kivell, et al. "*Homo naledi*, a New Species of the Genus *Homo* from the Dinaledi Chamber, South Africa." *eLife* 4 (2015): e09560.

Bischoff, James L., Narcís Soler, and Julià Maroto Ramon. "Abrupt Mousterian/Aurignacian Boundary at c. 40 ka BP: Accelerator ^{14}C Dates from l'Arbreda Cave (Catalunya, Spain)." *Journal of Archaeological Science* 16 (1989): 563–76.

Blumenbach, Johann Friedrich. *De Generis Humani Varietate Nativa*. 3rd ed. Göttingen: Vandenhoek and Ruprecht, 1795.

Bocherens, Hervé, Gennady Baryshnikov, and Wim Van Neer. "Were Bears or Lions Involved in Salmon Accumulation in the Middle Palaeolithic of the Caucasus? An Isotopic Investigation in Kudaro 3 Cave." *Quaternary International* 339–340 (2014): 112–18.

Bocherens, Hervé, Dorothée G. Drucker, Mietje Germonpré, Martina Lázničková-Galetová, Yuichi I. Naito, Christoph Wissing, Jaroslav Brůžek, and Martin Oliva. "Reconstruction of the Gravettian Food-Web at

Předmostí I Using Multi-Isotopic Tracking (^{13}C, ^{15}N, ^{34}S) of Bone Collagen." *Quaternary International* 359–360 (2015): 211–28.

Boddy, A. M., M. R. McGowen, C. C. Sherwood, L. I. Grossman, M. Goodman, and D. E. Wildman. "Comparative Analysis of Encephalization in Mammals Reveals Relaxed Constraints on Anthropoid Primate and Cetacean Brain Scaling." *Journal of Evolutionary Biology* 25 (2012): 981–94.

Bordes, Jean-Guillaume. "Les interstratifications Châtelperronien / Aurignacien du Roc-de-Combe et du Piage (Lot, France): Analyse taphonomique des industries lithiques; implications archéologiques." PhD diss., Université de Bordeaux I, 2002.

Boule, Marcellin. "L'homme fossile de la Chapelle-aux-Saints." *Annales de Paléontologie* 6 (1911): 111–72; 7 (1912): 21–56, 85–192; 8 (1913): 1–70.

——. "Sur l'âge des squelettes humains des grottes de Menton." *L'Anthropologie* 16 (1905): 506–15.

Bräuer, Günther. "A Craniological Approach to the Origin of Anatomically Modern *Homo sapiens* in Africa and Implications for the Appearance of Modern Humans." In *The Origins of Modern Humans: A World Survey of the Fossil Evidence*, ed. Fred H. Smith and Frank Spencer, 327–410. New York: Alan R. Liss, 1984.

Bricker, Harvey M. "The Provenience of Flint Used for the Manufacture of Tools." In *Excavation of the Abri Pataud (Les Eyzies, Dordogne)*, ed. Hallam L. Movius Jr., 194–98. Cambridge, MA: American School of Prehistoric Research, 1975.

Broca, Paul. "Sur les crânes et ossements des Eyzies." *Bulletins de la Société d'Anthropologie de Paris* 3 (1868): 350–92.

Burkitt, M. C. *The Old Stone Age: A Study of Palaeolithic Times*. New York: New York University Press, 1956.

Cabrera, Victoria Valdes, and James L. Bischoff. "Accelerator ^{14}C Dates for Early Upper Paleolithic (Basal Aurignacian) at El Castillo Cave (Spain)." *Journal of Archaeological Science* 16 (1989): 577–84.

Cai, Yanjun, Xiaoke Qiang, Xulong Wang, Changzhu Jin, Yuan Wang, Yingqi Zhang, Erik Trinkaus, and Zhisheng An. "The Age of Human Remains and Associated Fauna from Zhiren Cave in Guangxi, Southern China." *Quaternary International* 434 (2017): 84–91.

Cann, Rebecca L., Mark Stoneking, and Allan C. Wilson. "Mitochondrial DNA and Human Evolution." *Nature* 325 (1987): 31–36.

Caramelli, David, Carles Lalueza-Fox, Cristiano Vernesi, Martina Lari, Antonella Casoli, Francesco Mallegni, Brunetto Chiarelli, et al. "Evidence for a Genetic Discontinuity Between Neandertals and 24,000-Year-Old Anatomically Modern Europeans." *Proceedings of the National Academy of Sciences (USA)* 100 (2003): 6593–97.

Caron, François, Francesco d'Errico, Pierre Del Moral, Frédéric Santos, and João Zilhão. "The Reality of Neandertal Symbolic Behavior at the Grotte du Renne, Arcy-sur-Cure, France." *PLoS ONE* 6 (2011): e21545.

Cartailhac, Émile, and Henri Breuil. *La caverne d'Altamira à Santillane près Santander (Espagne)*. Monaco: Imprimerie de Monaco, 1906.

Cartailhac, Émile, and Eugène Trutat. "Sur la distinction à établir entre les races humaines dont on a trouvé les traces dans la grotte d'Aurignac." *Comptes Rendus de l'Académie des Sciences Paris* 73 (1871): 353–54.

Cartmill, Matt, and Fred H. Smith. *The Human Lineage*. 2nd ed. Hoboken, NJ: Wiley, 2022.

Charlier, Philippe, Nadia Benmoussa, Philippe Froesch, Isabelle Huynh-Charlier, and Antoine Balzeau. "Did Cro-Magnon 1 have Neurofibromatosis Type 1?" *The Lancet* 391 (2018): 1259.

Chen, Fahu, Frido Welker, Chuan-Chou Shen, Shara E. Bailey, Inga Bergmann, Simon Davis, Huan Xia, et al. "A Late Middle Pleistocene Denisovan Mandible from the Tibetan Plateau." *Nature* 569 (2019): 409–12.

Churchill, Steven E. *Thin on the Ground: Neandertal Biology, Archeology, and Ecology*. Hoboken, NJ: Wiley Blackwell, 2014.

——. "Weapon Technology, Prey Size Selection, and Hunting Methods in Modern Hunter-Gatherers: Implications for Hunting in the Palaeolithic and Mesolithic." In *Hunting and Animal Exploitation in the*

Later Palaeolithic and Mesolithic of Eurasia, ed. Gail L. Peterkin, Harvey Bricker, and Paul A. Mellars, 11–24. Washington, DC: American Anthropological Association, 1993.

Churchill Steven E., and Vincenzo Formicola. "A Case of Marked Bilateral Asymmetry in the Upper Limbs of an Upper Palaeolithic Male from Barma Grande (Liguria), Italy." *International Journal of Osteoarchaeology* 7 (1997): 18–38.

Claparède, Édouard. "Review of *Sur l'Origine des Espèces*, by Charles Darwin." *Revue Germanique*, October 1861.

Clark, Grahame. *World Prehistory: A New Outline*. Cambridge: Cambridge University Press, 1969.

Clark, J. Desmond, Yonas Beyene, Giday WoldeGabriel, William K. Hart, Paul R. Renne, Henry Gilbert, Alban Defleur, et al. "Stratigraphic, Chronological and Behavioural Contexts of Pleistocene *Homo sapiens* from Middle Awash, Ethiopia." *Nature* 423 (2003): 747–52.

Conard, Nicholas J., Maria Malina, and Susanne C. Münzel. "New Flutes Document the Earliest Musical Tradition in Southwestern Germany." *Nature* 460 (2009): 737–40.

Conservation Biology Institute. "Data Basin," accessed August 5, 2021, databasin.org.

Coon, Carleton S. *The Origin of Races*. New York: Knopf, 1962.

Cortés-Sánchez, Miguel, Francisco J. Jiménez-Espejo, María D. Simón-Vallejo, Chris Stringer, María Carmen Lozano Francisco, Antonio García-Alix, José L. Vera Peláez, et al. "An Early Aurignacian Arrival in Southwestern Europe." *Nature Ecology & Evolution* 3 (2019): 207–12.

Cristiani, Emanuela, Anita Radini, Marija Edinborough, and Dušan Borić. "Dental Calculus Reveals Mesolithic Foragers in the Balkans Consumed Domesticated Plant Foods." *Proceedings of the National Academy of Sciences (USA)* 113 (2016): 10298–303.

Cuvier, Georges. *Le Règne Animal*. Paris: A. Belin (Chez Déterville), 1817.

Darroch, J. N., and J. E. Mosimann. "Canonical and Principal Components of Shape." *Biometrika* 72 (1985): 241–52.

Darwin, Charles. *The Descent of Man, and Selection in Relation to Sex*. London: John Murray, 1871.

——. *On the Origin of Species by Means of Natural Selection*. London: John Murray, 1859.

Dastugue, Jean. "Pathologie des hommes fossiles de l'Abri de Cro-Magnon." *L'Anthropologie* 71 (1967): 479–92.

David, Francine, Nelly Connet, Michel Girard, Vincent Lhomme, Jean-Claude Miskovsky, and Annie Roblin-Jouve. "Le Châtelperronien de la Grotte du Renne à Arcy-sur-Cure (Yonne). Données sédimentologiques et chronostratigraphiques." *Bulletin de la Société Préhistorique Française* 98 (2001): 207–30.

Davidson, Iain. "The Colonization of Australia and Its Adjacent Islands and the Evolution of Modern Cognition." *Current Anthropology* 51, suppl. (2010): S177–89.

Day, Michael H., and Chris B. Stringer. "A Reconsideration of the Omo Kibish Remains and the *erectus–sapiens* Transition." In *L'*Homo erectus *et la Place de l'Homme de Tautavel Parmi les Hominides Fossiles*, ed. Henry De Lumley, 814–46. Nice: Centre National de la Recherche Scientifique, 1982.

D'Errico, Francesco. "The Invisible Frontier: A Multiple Species Model for the Origin of Behavioral Modernity." *Evolutionary Anthropology* 12 (2003): 188–202.

D'Errico, Francesco, João Zilhão, Michèle Julien, Dominique Baffier, and Jacques Pelegrin. "Neanderthal Acculturation in Western Europe? A Critical Review of the Evidence and Its Interpretation." *Current Anthropology* 39, suppl. (1998): S1–44.

Dibble, Harold L., Vera Aldeias, Paul Goldberg, Shannon P. McPherron, Dennis Sandgathe, and Teresa E. Steele. "A Critical Look at Evidence from La Chapelle-aux-Saints Supporting an Intentional Neandertal Burial." *Journal of Archaeological Science* 53 (2015): 649–57.

Diercke, Carl, Wiebke Gehring, and Thomas Michael. *Diercke Weltatlas*. Braunschweig, Germany: Westermann, 2008.

Dirks, Paul H. G. M., Eric M. Roberts, Hannah Hilbert-Wolf, Jan D. Kramers, John Hawks, Anthony Dosseto, Mathieu Duval, et al. "The Age of *Homo naledi* and Associated Sediments in the Rising Star Cave, South Africa." *eLife* 6 (2017): e24231.

Dodat, Pierre-Jean, Théo Tacail, Emmanuelle Albalat, Asier Gómez-Olivencia, Christine Couture-Veschambre, Trenton Holliday, Stéphane Madelaine, et al. "Isotopic Calcium Biogeochemistry of MIS 5 Fossil Vertebrate Bones: Application to the Study of the Dietary Reconstruction of Regourdou 1 Neandertal Fossil." *Journal of Human Evolution* 151 (2021): 102925.

Douka, Katerina, Stefano Grimaldi, Giovanni Boschian, Angiolo del Lucchese, and Thomas F. G. Higham. "A New Chronostratigraphic Framework for the Upper Palaeolithic of Riparo Mochi (Italy)." *Journal of Human Evolution* 62 (2012): 286–99.

Duarte, Cidália, João Maurício, Paul B. Pettitt, Pedro Souto, Erik Trinkaus, Hans van der Plicht, and João Zilhão. "The Early Upper Paleolithic Human Skeleton from the Abrigo do Lagar Velho (Portugal) and Modern Human Emergence in Iberia." *Proceedings of the National Academy of Sciences (USA)* 96 (1999): 7604–9.

Eren, Metin I. "Comment to Shea." *Current Anthropology* 52 (2011): 18.

Evans, Arthur J. "On the Prehistoric Interments of the Balzi Rossi Caves Near Mentone and Their Relation to the Neolithic Cave-Burials of the Finalese." *Journal of the Anthropological Institute of Great Britain and Ireland* 22 (1893): 286–307.

Falcucci, Armando, Nicholas J. Conard, and Marco Peresani. "A Critical Assessment of the Protoaurignacian Lithic Technology at Fumane Cave and Its Implications for the Definition of the Earliest Aurignacian." *PLoS ONE* 12 (2017): e0189241.

Fernández, Santiago, Noemí Fuentes, José S. Carrión, Penélope González-Sampériz, Encarna Montoya, Graciela Gil, Gerardo Vega-Toscano, and José A. Riquelme. "The Holocene and Upper Pleistocene Pollen Sequence of Carihuela Cave, Southern Spain." *Geobios* 40 (2007): 75–90.

Fewlass, Helen, Sahra Talamo, Lukas Wacker, Bernd Kromer, Thibaut Tuna, Yoann Fagault, Edouard Bard, et al. "A ¹⁴C Chronology for the Middle to Upper Palaeolithic Transition at Bacho Kiro Cave, Bulgaria." *Nature Ecology & Evolution* 4 (2020): 794–801.

Finlayson, Clive. *Neanderthals and Modern Humans: An Ecological and Evolutionary Perspective.* Cambridge: Cambridge University Press, 2004.

Fischer, Anne, Victor Wiebe, Svante Pääbo, and Molly Przeworski. "Evidence for a Complex Demographic History of Chimpanzees." *Molecular Biology and Evolution* 21 (2004): 799–808.

Fischer, Paul. "Sur la conchyliologie des cavernes." *Bulletins de la Société d'Anthropologie de Paris*, Deuxième série, 11 (1876): 181–85.

Fleagle, John G., Zelalem Assefa, Francis H. Brown, and John J. Shea. "Paleoanthropology of the Kibish Formation, Southern Ethiopia: Introduction." *Journal of Human Evolution* 55 (2008): 360–65.

Formicola, Vincenzo, and Brigitte M. Holt. "Tall Guys and Fat Ladies: Grimaldi's Upper Paleolithic Burials and Figurines in an Historical Perspective." *Journal of Anthropological Sciences* 93 (2015): 1–18.

Fraipont, Julien, and Max Lohest. "La race humaine de Néanderthal ou de Canstadt en Belgique: Recherches ethnographiques sur des ossements humains, découvertes dans les dépôts quaternaires d'une grotte à Spy et détermination de leur âge géologique." *Archives de Biologie* 7 (1887): 587–757.

Franciscus, Robert G., and Trenton W. Holliday. "Hindlimb Skeletal Allometry in Plio-Pleistocene Hominids with Special Reference to AL 288-1 ('Lucy')." *Bulletins et Mémoires de la Société d'Anthropologie de Paris* n.s., 4 (1992): 5–20.

Franciscus Robert G., and Erik Trinkaus. "Determinants of retromolar space presence in Pleistocene *Homo* mandibles." *Journal of Human Evolution* 28 (1995): 577–95.

Fu, Qiaomei, Mateja Hajdinjak, Oana Teodora Moldovan, Silviu Constantin, Swapan Mallick, Pontus Skoglund, Nick Patterson, et al. "An Early Modern Human from Romania with a Recent Neanderthal Ancestor." *Nature* 524 (2015): 216–19.

Fu, Qiaomei, Heng Li, Priya Moorjani, Flora Jay, Sergey M. Slepchenko, Aleksei A. Bondarev, Philip L. F. Johnson, et al. "Genome Sequence of a 45,000-Year-Old Modern Human from Western Siberia." *Nature* 514 (2014): 445–49.

Fu, Qiaomei, Cosimo Posth, Mateja Hajdinjak, Martin Petr, Swapan Mallick, Daniel Fernandes, Anja Furtwängler, et al. "The Genetic History of Ice Age Europe." *Nature* 534 (2016): 200–205.

García-González, Rebeca, José Miguel Carretero, Michael P. Richards, Laura Rodríguez, and Rolf Quam. "Dietary Inferences Through Dental Microwear and Isotope Analyses of the Lower Magdalenian Individual from El Mirón Cave (Cantabria, Spain)." *Journal of Archaeological Science* 60: (2015): 28–38.

Garn, Stanley M. *Human Races*. Springfield, IL: Charles C Thomas, 1961.

Germonpré, Mietje, Martina Lázničková-Galetová, Robert J. Losey, Jannikke Räikkönen, and Mikhail V. Sablin. "Large Canids at the Gravettian Předmostí Site, the Czech Republic: The Mandible." *Quaternary International* 359–360 (2015) 261–79.

Germonpré, Mietje, Martina Lázničková-Galetová, and Mikhail V. Sablin. "Palaeolithic Dog Skulls at the Gravettian Předmostí Site, the Czech Republic." *Journal of Archaeological Science* 39 (2012): 184–202.

Gravina, Brad, François Bachellerie, Solène Caux, Emmanuel Discamps, Jean-Philippe Faivre, Aline Galland, Alexandre Michel, Nicolas Teyssandier, and Jean-Guillaume Bordes. "No Reliable Evidence for a Neanderthal-Châtelperronian Association at La Roche-à-Pierrot, Saint-Césaire." *Nature Scientific Reports* 8 (2018): 15134.

Green, Richard E., Johannes Krause, Adrian W. Briggs, Tomislav Maricic, Udo Stenzel, Martin Kircher, Nick Patterson, et al. "A Draft Sequence of the Neandertal Genome." *Science* 328 (2010): 710–22.

Green, Richard E., Johannes Krause, Susan E. Ptak, Adrian W. Briggs, Michael T. Ronan, Jan F. Simons, Lei Du, et al. "Analysis of One Million Base Pairs of Neanderthal DNA." *Nature* 444 (2006): 330–36.

Grün, Rainer, James S. Brink, Nigel A. Spooner, Chris B. Stringer, Robert G. Franciscus, and Andrew S. Murray. "Direct Dating of Florisbad Hominid." *Nature* 382 (1996): 500–501.

Hajdinjak, Mateja, Fabrizio Mafessoni, Laurits Skov, Benjamin Vernot, Alexander Hübner, Qiaomei Fu, Elena Essel, et al. "Initial Upper Palaeolithic Humans in Europe Had Recent Neanderthal Ancestry." *Nature* 592 (2021): 253–57.

Hardy, Karen, Stephen Buckley, and Les Copeland. "Pleistocene Dental Calculus: Recovering Information on Paleolithic Food Items, Medicines, Paleoenvironment and Microbes." *Evolutionary Anthropology* 27 (2018): 234–46.

Harmand, Sonia, Jason E. Lewis, Craig S. Feibel, Christopher J. Lepre, Sandrine Prat, Arnaud Lenoble, Xavier Boës, et al. "3.3-Million-Year-Old Stone Tools from Lomekwi 3, West Turkana, Kenya." *Nature* 521 (2015): 310–15.

Harvati, Katerina, Carolin Röding, Abel M. Bosman, Fotios A. Karakostis, Rainer Grün, Chris Stringer, Panagiotis Karkanas, et al. "Apidima Cave Fossils Provide Earliest Evidence of *Homo sapiens* in Eurasia." *Nature* 571 (2019): 500–504.

Haws, Jonathan A., Michael M. Benedetti, Sahra Talamo, Nuno Bicho, João Cascalheira, M. Grace Ellis, Milena M. Carvalho, Lukas Friedl, Telmo Pereira, and Brandon K. Zinsious. "The Early Aurignacian Dispersal of Modern Humans into Westernmost Eurasia." *Proceedings of the National Academy of Sciences (USA)* 117 (2020): 25414–22.

Henry-Gambier, Dominique. "Les fossiles de Cro-Magnon (Les Eyzies-de-Tayac, Dordogne) Nouvelles données sur leur position chronologique et leur attribution culturelle." *Bulletins et Mémoires de la Société d'Anthropologie de Paris* 14 (2002): 89–112.

Henshilwood, Christopher S., and Curtis W. Marean. "The Origin of Modern Human Behavior: Critique of the Models and Their Test Implications." *Current Anthropology* 44 (2003): 627–51.

Hershkovitz, Israel, Gerhard W. Weber, Rolf Quam, Mathieu Duval, Rainer Grün, Leslie Kinsley, Avner Ayalon, et al. "The Earliest Modern Humans Outside Africa." *Science* 359 (2018): 456–59.

Higham, Thomas, Laura Basell, Roger Jacobi, Rachel Wood, Christopher Bronk Ramsey, and Nicholas J. Conard. "Testing Models for the Beginnings of the Aurignacian and the Advent of Figurative Art and Music: The Radiocarbon Chronology of Geißenklösterle." *Journal of Human Evolution* 62 (2012): 664–76.

Hillson, Simon W., Robert G. Franciscus, Trenton W. Holliday, and Erik Trinkaus. "The Ages at Death." In *Early Modern Human Evolution in Central Europe: The People of Dolní Věstonice and Pavlov*, ed. Erik Trinkaus and Jiří Svoboda, 31–45. Oxford: Oxford University Press, 2006.

Hoffmann, Dirk L., Diego E. Angelucci, Valentín Villaverde, Josefina Zapata, and João Zilhão. "Symbolic Use of Marine Shells and Mineral Pigments by Iberian Neandertals 115,000 Years Ago." *Science Advances* 4 (2018): eaar5255.

Hoffmann, D. L., C. D. Standish, M. García-Diez, P. B. Pettitt, J. A. Milton, J. Zilhão, J. J. Alcolea-González, et al. "U-Th Dating of Carbonate Crusts Reveals Neandertal Origin of Iberian Cave Art." *Science* 359 (2018): 912–15.

Holliday, Trenton W. "Body Proportions in Late Pleistocene Europe and Modern Human Origins." *Journal of Human Evolution* 32 (1997): 423–48.

——. "Body Size and Proportions in the Late Pleistocene Western Old World and the Origins of Modern Humans." PhD diss., University of New Mexico, 1995.

——. "Comment to Henshilwood and Marean." *Current Anthropology* 44 (2003): 639–40.

——. "The Ecological Context of Trapping Among Recent Hunter-Gatherers: Implications for Subsistence in Terminal Pleistocene Europe." *Current Anthropology* 39 (1998): 711–20.

Holliday, Trenton W., and Anthony B. Falsetti. "A New Method for Discriminating African-American from European-American Skeletons Using Postcranial Osteometrics Reflective of Body Shape." *Journal of Forensic Sciences* 44 (1999): 926–30.

Holliday, Trenton W., and Charles E. Hilton. "Body Proportions of Circumpolar Peoples as Evidenced from Skeletal Data: Ipiutak and Tigara (Point Hope) Versus Kodiak Island Inuit." *American Journal of Physical Anthropology* 29 (2010): 287–302.

Hooton, Earnest Albert. "Methods of Racial Analysis." *Science,* n.s., 63 (1926): 75–81.

——. *Up From the Ape.* New York: Macmillan, 1935.

Howells, William W. *Cranial Variation in Man. A Study by Multivariate Analysis of Patterns of Differences Among Recent Human Populations.* Papers of the Peabody Museum of Archeology and Ethnology. vol. 67. Cambridge, MA: Peabody Museum, 1973.

Hublin, Jean-Jacques, Abdelouahed Ben-Ncer, Shara E. Bailey, Sarah E. Freidline, Simon Neubauer, Matthew M. Skinner, Inga Bergmann, et al. "New Fossils from Jebel Irhoud, Morocco and the Pan-African Origin of *Homo sapiens.*" *Nature* 546 (2017): 290–92.

Hublin, Jean-Jacques, Cecilio Barroso Ruiz, Paqui Medina Lara, Michel Fontugne, and Jean-Louis Reyss. "The Mousterian Site of Zafarraya (Andalucia, Spain): Dating and Implications on the Palaeolithic Peopling Processes of Western Europe." *Comptes Rendus de l'Academie de Sciences II* 32 (1995): 931–37.

Hublin, Jean-Jacques, Nikolay Sirakov, Vera Aldeias, Shara Bailey, Edouard Bard, Vincent Delvigne, Elena Endarova, et al. "Initial Upper Palaeolithic *Homo sapiens* from Bacho Kiro Cave, Bulgaria." *Nature* 581 (2020): 299–302.

Hublin, Jean-Jacques, Fred Spoor, Marc Braun, Frans Zonneveld, and Silvana Condemi. "A Late Neanderthal Associated with Upper Paleolithic Artefacts." *Nature* 381 (1996): 224–26.

Hublin, Jean-Jacques, Sahra Talamo, Michèle Julien, Francine David, Nelly Connet, Pierre Bodu, Bernard Vandermeersch, and Michael P. Richards. "Radiocarbon Dates from the Grotte du Renne and Saint-Césaire Support a Neandertal Origin for the Châtelperronian." *Proceedings of the National Academy of Sciences (USA)* 109 (2012): 18743–48.

Huerta-Sánchez, Emilia, Xin Jin, Asan, Zhuoma Bianba, Benjamin M. Peter, Nicolas Vinckenbosch, Yu Liang, et al. "Altitude Adaptation in Tibetans Caused by Introgression of Denisovan-like DNA." *Nature* 512 (2014): 194–97.

Hughes, Susan S. "Getting to the Point: Evolutionary Change in Prehistoric Weaponry." *Journal of Archaeological Method and Theory* 5 (1998): 345–408.

Hurel, A. "1868: Le moment Cro-Magnon." *Bulletins et Mémoires de la Société d'Anthropologie de Paris* 30 (2018): 111–20.

Huxley, Thomas H. *Evidence as to Man's Place in Nature.* London: Williams and Norgate, 1863.

Iakovleva, Lioudmila. "L'art mézinien en Europe orientale dans son context chronologique, culturel et spirituel." *L'Anthropologie* 113 (2009): 691–752.

Jacobi, R. M., and T. F. G. Higham. "The 'Red Lady' Ages Gracefully: New Ultrafiltration AMS Determinations from Paviland." *Journal of Human Evolution* 55 (2008): 898–907.

Jaubert, Jacques, Sophie Verheyden, Dominique Genty, Michel Soulier, Hai Cheng, Dominique Blamart, Christian Burlet, et al. "Early Neanderthal Constructions Deep in Bruniquel Cave in Southwestern France." *Nature* 534 (2016): 111–14.

Jerison, Harry J. *Evolution of the Brain and Intelligence*. New York: Academic Press, 1973.

Ji, Qiang, Wensheng Wu, Yannan Ji, Qiang Li, and Xijun Ni. "Late Middle Pleistocene Harbin Cranium Represents New *Homo* Species." *Innovation* 2 (2021): 100132.

Jordan, Peter A., Daniel B. Botkin, and Michael L. Wolfe. "Biomass Dynamics in a Moose Population." *Ecology* 52 (1971): 147–52.

Jørgensen-Rideout, Sophie. "Hot Rocks, Cold Data: An Examination of Fire Use at the Upper Palaeolithic Site of Abri Pataud." RMA thesis, University of Leiden, 2019.

Karavanić, Ivor. "Olschewian and Appearance of Bone Technology in Croatia and Slovenia." In *Neanderthals and Modern Humans—Discussing the Transition. Central and Eastern Europe from 50.000–30.000 B.P.*, ed. Jorg Örschiedt and Gerd-Christian Weniger, 159–68. Mettmann, Neandertal Museum, 2000.

Kasabova, Boryana E., and Trenton W. Holliday. "New Model for Estimating the Relationship Between Surface Area and Volume in the Human Body Using Skeletal Remains." *American Journal of Physical Anthropology* 156 (2015): 614–24.

Keesing, Felicia. "Cryptic Consumers and the Ecology of an African Savanna." *BioScience* 50 (2000): 205–15.

Kehl, Martin, Christoph Burow, Alexandra Hilgers, Marta Navazo, Andreas Pastoors, Gerd-Christian Weniger, Rachel Wood, and Jesús F. Jordá Pardo. "Late Neanderthals at Jarama VI (Central Iberia)?" *Quaternary Research* 80 (2013): 218–34.

Kidder, James H., Richard L. Jantz, and Fred H. Smith. "Defining Modern Humans: A Multivariate Approach." In *Continuity or Replacement: Controversies in* Homo sapiens *Evolution*, ed. Günter Bräuer and Fred H. Smith, 157–77. Rotterdam: A. A. Balkema, 1992.

King, William. "The Reputed Fossil Man of Neanderthal." *Quarterly Journal of Science* 1 (1864): 88–97.

Klein, Richard G. "Archeology and the Evolution of Human Behavior." *Evolutionary Anthropology* 9 (2000): 17–36.

Krause, Johannes, Qiaomei Fu, Jeffrey M. Good, Bence Viola, Michael V. Shunkov, Anatoli P. Derevianko, and Svante Pääbo. "The Complete Mitochondrial DNA Genome of an Unknown Hominin from Southern Siberia." *Nature* 464 (2010): 894–97.

Krause, Johannes, Ludovic Orlando, David Serre, Bence Viola, Kay Prüfer, Michael P. Richards, Jean-Jacques Hublin, Catherine Hänni, Anatoly P. Derevianko, and Svante Pääbo. "Neanderthals in Central Asia and Siberia." *Nature* 449 (2007): 902–4.

Krings, Matthias, Cristian Capelli, Frank Tschentscher, Helga Geisert, Sonja Meyer, Arndt von Haeseler, Karl Grossschmidt, Göran Possnert, Maja Paunovic, and Svante Pääbo. "A View of Neandertal Genetic Diversity." *Nature Genetics* 26 (2000): 144–46.

Krings, Matthias, Anne Stone, Ralf W. Schmitz, Heike Krainitzki, Mark Stoneking, and Svante Pääbo. "Neandertal DNA Sequences and the Origin of Modern Humans." *Cell* 90 (1997): 19–30.

Kroeber, Alfred L. "Sub-human Cultural Beginnings." *Quarterly Review of Biology* 3 (1928): 325–42.

Langley, Michelle C. "Late Pleistocene Osseous Projectile Technology and Cultural Variability." In *Osseous Projectile Weaponry: Towards an Understanding of Pleistocene Cultural Variability*, ed. Michelle C. Langley, 1–11. Dordrecht: Springer, 2016.

Lartet, Édouard. "Nouvelles recherches sur la coexistence de l'Homme et des grands mammifères fossiles réputés caractéristiques de la dernière période géologique." *Annales des Sciences Naturelles, Zoologie et Paléontologie* 15 (1861): 177–253.

Lartet, Louis. "Une sépulture des troglodytes du Périgord." *Bulletins de la Société d'Anthropologie de Paris* t. 3 (1868): 335–49.

Leder, Dirk, Raphael Hermann, Matthias Hüls, Gabriele Russo, Philipp Hoelzmann, Ralf Nielbock, Utz Böhner, et al. "A 51,000-Year-Old Engraved Bone Reveals Neanderthals' Capacity for Symbolic Behaviour." *Nature Ecology & Evolution* 5 (2021): 1273–82.

Lewontin, R. C. "The Apportionment of Human Diversity." In *Evolutionary Biology*, ed. Theodosius Dobzhansky, Max K. Hecht, and William C. Steere, 381–98. New York: Springer, 1972.

Linnaeus, Carolus. *Systema Naturae*. 10th ed. Stockholm: Laurentius Salvius, 1758.

Liu, Wu, María Martinón-Torres, Yan-jun Cai, Song Xing, Hao-wen Tong, Shu-wen Pei, Mark Jan Sier, et al. "The Earliest Unequivocally Modern Humans in Southern China." *Nature* 526 (2015): 696–700.

Lombard, Marlize. "Quartz-Tipped Arrows Older Than 60 ka: Further Use-Trace Evidence from Sibudu, KwaZulu-Natal, South Africa." *Journal of Archaeological Science* 38 (2011): 1918–30.

Long, Jeffrey C., Jie Li, and Meghan E. Healy. "Human DNA Sequences: More Variation and Less Race." *American Journal of Physical Anthropology* 139 (2009): 23–34.

Lumley, Henry de, ed. *La Grotte du Cavillon sous la falaise des Baousse Rousse, Grimaldi, Vintimille, Italie (Archéologie/Préhistoire)*. Paris: CNRS Éditions, 2016.

Lyell, Charles. *Principles of Geology*. 6th ed. London: John Murray, 1840.

Maddison, David R. "African Origin of Human Mitochondrial DNA Reexamined." *Systematic Zoology* 40 (1991): 355–63.

——. "The Discovery and Importance of Multiple Island Trees." *Systematic Zoology* 40 (1991): 315–28.

Maddux, Scott D., and Robert G. Franciscus. "Allometric Scaling of Infraorbital Surface Topography in *Homo*." *Journal of Human Evolution* 56 (2009): 161–74.

Manega, Paul C. "Geochronology, Geochemistry, and Isotopic Study of the Plio-Pleistocene Hominid Sites and the Ngorongoro Volcanic Highland in Northern Tanzania." PhD diss., University of Colorado, 1993.

Mann, Alan E. *World Book Encyclopedia*. 57th ed. s.v. "Prehistoric Man." Chicago: World Book, 1976.

Martin, Robert D. "Relative Brain Size and Basal Metabolic Rate in Terrestrial Vertebrates." *Nature* 293 (1981): 57–60.

Mayr, Ernst. *Principles of Systematic Zoology*. New York: McGraw-Hill, 1969.

McBrearty, Sally. "Comment to Henshilwood and Marean." *Current Anthropology* 44 (2003): 641–42.

McBrearty, Sally, and Allison S. Brooks. "The Revolution That Wasn't: A New Interpretation of the Origin of Modern Human Behavior." *Journal of Human Evolution* 39 (2000): 453–563.

McGrew, William C. *Chimpanzee Material Culture: Implications for Human Evolution*. Cambridge: Cambridge University Press, 1992.

Mellars, Paul. "The Impossible Coincidence. A Single-Species Model for the Origins of Modern Human Behavior in Europe." *Evolutionary Anthropology* 14 (2005): 12–27.

Mellars, Paul, and Brad Gravina. "Châtelperron: Theoretical Agendas, Archaeological Facts, and Diversionary Smoke-Screens." *PaleoAnthropology* (2008): 43–64.

Mensan, Romain, Raphaëlle Bourrillon, Catherine Cretin, Randall White, Philippe Gardère, Laurent Chiotti, Matthew Sisk, Amy Clark, Thomas Higham, and Élise Tartar. "Une nouvelle découverte d'art parietal aurignacien *in situ* à l'abri Castanet (Dordogne, France): Context et datation." *Paleo* 23 (2012): 171–88.

Mercier, N., H. Valladas, O. Bar-Yosef, B. Vandermeersch, C. Stringer, and J.-L. Joron. "Thermoluminescence Date for the Mousterian Burial Site of Es-Skhul, Mt. Carmel." *Journal of Archaeological Science* 20 (1993): 169–74.

Meyer, Matthias, Juan-Luis Arsuaga, Cesare de Filippo, Sarah Nagel, Ayinuer Aximu-Petri, Birgit Nickel, Ignacio Martínez, et al. "Nuclear DNA Sequences from the Middle Pleistocene Sima de los Huesos Hominins." *Nature* 531 (2016): 504–7.

Mittnik, Alissa, Chuan-Chao Wang, Jiří Svoboda, and Johannes Krause. "A Molecular Approach to the Sexing of the Triple Burial at the Upper Paleolithic Site of Dolní Věstonice." *PLoS ONE* 11 (2016): e0163019.

Montes, Ramón, Juan Sanguino, P. Martín, Antonio José Gómez, and C. Morcillo. "La secuencia estratigráfica de la cueva de El Pendo (Escobedo de Camargo, Cantabria): Problemas geoarqueológicos de un referente cronocultural." In *Geoarqueología y patrimonio en la Península Ibérica y el entorno mediterráneo*, ed. Manuel Santonja, Alfredo Pérez-González, and María José Machado, 139–59. Almazán: ADEMA, 2005.

Morales, Juan I., Artur Cebrià, Aitor Burguet-Coca, Juan Luis Fernández-Marchena, Gala García-Argudo, Antonio Rodríguez-Hidalgo, María Soto, et al. "The Middle-to-Upper Paleolithic Transition Occupations from Cova Foradada (Calafell, NE Iberia)." *PLoS ONE* 14, no. 5 (2019): e0215832.

Morant, G. M. "Studies of Palaeolithic Man IV. A Biometric Study of the Upper Palaeolithic Skulls of Europe and Their Relationships to Earlier and Later Types." *Annals of Eugenics* 4 (1930): 109–214.

Morin, Eugène, and Véronique Laroulandie. "Presumed Symbolic Use of Diurnal Raptors by Neanderthals." *PLoS ONE* 7 (2012): e32856.

Mortillet, Gabrielle de. "Classification des diverses périodes de l'Âge de la pierre." *Congrès International d'Anthropologie et d'Archéologie Préhistorique, Bruxelles* (1871): 432–59.

Nerudová, Zdeňka, Eva Vaníčková, Zdeněk Tvrdý, Jiří Ramba, Ondřej Bílek, and Petr Kostrhun. "The Woman from the Dolní Věstonice 3 Burial: A New View of the Face Using Modern Technologies." *Archaeological and Anthropological Sciences* 11 (2019): 2527–38.

Niedbalski, Sara D., and Jeffrey C. Long. "Novel Alleles Gained During the Beringian Isolation Period." *Nature Scientific Reports* 12 (2022): 4289.

Nigst, Philip R., Paul Haesaerts, Freddy Damblon, Christa Frank-Fellner, Carolina Mallol, Bence Viola, Michael Götzinger, Laura Niven, Gerhard Trnka, and Jean-Jacques Hublin. "Early Modern Human Settlement of Europe North of the Alps Occurred 43,500 Years Ago in a Cold Steppe-Type Environment." *Proceedings of the National Academy of Sciences (USA)* 111 (2014): 14394–99.

Norton, Heather L., Rick A. Kittles, Esteban Parra, Paul McKeigue, Xianyun Mao, Keith Cheng, Victor A. Canfield, Daniel G. Bradley, Brian McEvoy, and Mark D. Shriver. "Genetic Evidence for the Convergent Evolution of Light Skin in Europeans and East Asians." *Molecular Biology and Evolution* 24 (2007): 710–22.

Nowell, April. "Comment to Shea." *Current Anthropology* 52 (2011): 19–20.

——. "Defining Behavioral Modernity in the Context of Neandertal and Anatomically Modern Human Populations." *Annual Review of Anthropology* 39 (2010): 437–52.

——. *Growing Up in the Ice Age: Fossil and Archaeological Evidence of the Lived Lives of Plio-Pleistocene Children*. Oxford: Oxbow Books, 2021.

Ovchinnikov, I. V., A. Götherström, G. P. Romanova, V. M. Kharitonov, K. Lidén, and W. Goodwin. "Molecular Analysis of Neanderthal DNA from the Northern Caucasus." *Nature* 404 (2000): 490–93.

Pääbo, Svante. "Molecular Cloning of Ancient Egyptian Mmummy DNA." *Nature* 314 (1985): 644–45.

Partiot, Caroline, Erik Trinkaus, Christopher J. Knüsel, and Sébastien Villotte. "The Cro-Magnon Babies: Morphology and Mortuary Implications of the Cro-Magnon Immature Remains." *Journal of Archaeological Science: Reports* 30 (2020): 102257.

Peresani, Marco, Ivana Fiore, Monica Gala, Matteo Romandini, and Antonio Tagliacozzo. "Late Neandertals and the Intentional Removal of Feathers as Evidenced from Bird Bone Taphonomy at Fumane Cave 44 ky B.P., Italy." *Proceedings of the National Academy of Sciences (USA)* 108 (2011): 3888–93.

Pettitt, Paul, and Paul Bahn. "An Alternative Chronology for the Art of Chauvet Cave." *Antiquity* 89 (2015): 542–53.

Pike, A. W. G., D. L. Hoffmann, M. García-Diez, P. B. Pettitt, J. Alcolea, R. De Balbín, C. González-Sainz, et al. "U-Series Dating of Paleolithic Art in 11 Caves in Spain." *Science* 336 (2012): 1409–13.

Prüfer, Kay, Cosimo Posth, He Yu, Alexander Stoessel, Maria A. Spyrou, Thibaut Deviese, Marco Mattonai, et al. "A Genome Sequence from a Modern Human Skull over 45,000 Years Old from Zlatý kůň in Czechia." *Nature Ecology & Evolution* 5 (2021): 820–82.

Prüfer, Kay, Fernando Racimo, Nick Patterson, Flora Jay, Sriram Sankararaman, Susanna Sawyer, Anja Heinze, et al. "The Complete Genome Sequence of a Neanderthal from the Altai Mountains." *Nature* 504 (2014): 43–49.

Quiles, Anita, Hélène Valladas, Hervé Bocheren, Emmanuelle Delqué-Količ, Evelyne Kaltnecker, Johannes van der Plicht, Jean-Jacques Delannoy, et al. "A High Precision Chronological Model for the Decorated Upper Paleolithic Cave of Chauvet-Pont d'Arc, Ardèche, France." *Proceedings of the National Academy of Sciences (USA)* 113 (2016): 4670–75.

Radovčić, Davorka, Giovanni Birarda, Ankica Oros Sršen, Lisa Vaccari, Jakov Radovčić, and David W. Frayer. "Surface Analysis of an Eagle Talon from Krapina." *Nature Scientific Reports* 10 (2020): 6329.

Reich, David, Richard E. Green, Martin Kircher, Johannes Krause, Nick Patterson, Eric Y. Durand, Bence Viola, et al. "Genetic History of an Archaic Hominin Group from Denisova Cave in Siberia." *Nature* 468 (2010): 1054–59.

Rendu, William, Cédric Beauval, Isabelle Crevecoeur, Priscilla Bayle, Antoine Balzeau, Thierry Bismuth, Laurence Bourguignon, et al. "Let the Dead Speak... Comments on Dibble et al.'s Reply to 'Evidence Supporting an Intentional Burial at La Chapelle-aux-Saints.'" *Journal of Archaeological Science* 69 (2016): 12–20.

Rendu, William, Cédric Beauval, Isabelle Crevecoeur, Priscilla Bayle, Antoine Balzeau, Thierry Bismuth, Laurence Bourguignon, et al. "Evidence Supporting an Intentional Neandertal Burial at La Chapelle-aux-Saints." *Proceedings of the National Academy of Sciences (USA)* 111 (2014): 81–86.

Renne, P. R., D. Rennew, L. Sharpa, O. Deinog, and L. Civetta. "^{40}Ar/^{39}Ar Dating into the Historical Realm: Calibration Against Pliny the Younger." *Science* 277 (1997): 1279–80.

Revedin, Anna, Laura Longo, Marta Mariotti Lippi, Emanuele Marconi, Annamaria Ronchitelli, Jiri Svoboda, Eva Anichini, Matilde Gennai, and Biancamaria Aranguren. "New Technologies for Plant Food Processing in the Gravettian." *Quaternary International* 359–360 (2015): 77–88.

Richards, Michael P., Paul B. Pettitt, Mary C. Stiner, and Erik Trinkaus. "Stable Isotope Evidence for Increasing Dietary Breadth in the European Mid-Upper Paleolithic." *Proceedings of the National Academy of Sciences (USA)* 98 (2001): 6528–32.

Richter, Daniel, Rainer Grün, Renaud Joannes-Boyau, Teresa E. Steele, Fethi Amani, Mathieu Rué, Paul Fernandes, et al. "The Age of the Hominin Fossils from Jebel Irhoud, Morocco, and the Origins of the Middle Stone Age." *Nature* 546 (2017): 293–96.

Riel-Salvatore, Julien, Alexandra E. Miller, and Geoffrey A. Clark. "An Empirical Evaluation of the Case for a Châtelperronian-Aurignacian Interstratification at Grotte des Fées de Châtelperron." *World Archaeology* 40 (2008): 480–92.

Roberts, Derek F. *Climate and Human Variability*. 2nd ed. Menlo Park, CA: Cummings, 1978.

Rocca, Roxane, Nelly Connet, and Vincent Lhomme. "Before the Transition? The Final Middle Palaeolithic Lithic Industry from the Grotte du Renne (Layer XI) at Arcy-sur-Cure (Burgundy, France)." *Comptes Rendus Palevol* 16 (2017): 878–93.

Rolandi, G., Paone, A., Di Lascio, M., and G. Stefani. "The 79 AD Eruption of Somma: The Relationship Between the Date of the Eruption and the Southeast Tephra Dispersion." *Journal of Volcanology and Geothermal Research* 169 (2007): 87–98.

Roth, Gerhard, and Ursula Dicke. "Evolution of the Brain and Intelligence." *Trends in Cognitive Sciences* 9 (2005): 250–57.

Ruff, Christopher B., Erik Trinkaus, and Trenton W. Holliday. "Body Mass and Encephalization in Pleistocene *Homo*." *Nature* 387 (1997): 173–76.

Sano, Katsuhiro, Simona Arrighi, Chiaramaria Stani, Daniele Aureli, Francesco Boschin, Ivana Fiore, Vincenzo Spagnolo, et al. "The Earliest Evidence for Mechanically Delivered Projectile Weapons in Europe." *Nature Ecology & Evolution* 3 (2019): 1409–14.

Schmitz, Ralf W., David Serre, Georges Bonani, Susanne Feine, Felix Hillgruber, Heike Krainitzki, Svante Pääbo, and Fred H. Smith. "The Neandertal Type Site Revisited: Interdisciplinary Investigations of Skeletal Remains from the Neander Valley, Germany." *Proceedings of the National Academy of Sciences (USA)* 99 (2002): 13342–47.

Serrat, Maria A. "Allen's Rule Revisited: Temperature Influences Bone Elongation During a Critical Period of Postnatal Development." *Anatomical Record* 296 (2013): 1534–45.

Serrat, M. A., R. M. Williams, and C. E. Farnum. "Temperature Alters Solute Transport in Growth Plate Cartilage Measured by In Vivo Multiphoton Microscopy." *Journal of Applied Physiology* 106 (2009): 2016–25.

Serre, David, André Langaney, Mario Chech, Maria Teschler-Nicola, Maja Paunovic, Philippe Mennecier, Michael Hofreiter, Göran Possnert, and Svante Pääbo. "No Evidence of Neandertal mtDNA Contribution to Early Modern Humans." *PLoS Biology* 2 (2004): E57.

Shea, John J. "*Homo sapiens* Is as *Homo sapiens* Was: Behavioral Variability Versus 'Behavioral Modernity' in Paleolithic Archaeology." *Current Anthropology* 52 (2011): 1–35.

——. "The Origins of Lithic Projectile Point Technology: Evidence from Africa, the Levant, and Europe." *Journal of Archaeological Science* 33 (2006): 823–46.

——. *Stone Tools in Human Evolution: Behavioral Differences Among Technological Primates.* Cambridge: Cambridge University Press, 2017.

Shipman, Pat. *The Invaders: How Humans and Their Dogs Drove Neanderthals to Extinction.* Cambridge, MA: Belknap Press of Harvard University Press, 2015.

Slimak, Ludovic, Clément Zanolli, Tom Higham, Marine Frouin, Jean-Luc Schwenninger, Lee J. Arnold, Martina Demuro, et al. "Modern Human Incursion into Neanderthal Territories 54,000 Years Ago at Mandrin, France." *Science Advances* 8 (2022): eabj9496.

Slon, Viviane, Fabrizio Mafessoni, Benjamin Vernot, Cesare de Filippo, Steffi Grote, Bence Viola, Mateja Hajdinjak, et al. "The Genome of the Offspring of a Neanderthal Mother and a Denisovan Father." *Nature* 561 (2018): 7721.

Smith, Fred H. "Fossil Hominids from the Upper Pleistocene of Central Europe and the Origin of Modern Europeans." In *The Origins of Modern Humans: A World Survey of the Fossil Evidence*, ed. Fred H. Smith and Frank Spencer, 137–209. New York: Alan R. Liss, 1984.

Smith, Fred H., and Frank Spencer, eds. *The Origins of Modern Humans: A World Survey of the Fossil Evidence.* New York: Alan R. Liss, 1984.

Sollas, William J. "Paviland Cave: An Aurignacian Station in Wales." *Journal of the Royal Anthropological Institute* 43 (1913): 325–74.

Sparacello, Vitale S., Sébastien Villotte, Laura L. Shackelford, and Erik Trinkaus. "Patterns of Humeral Asymmetry Among Late Pleistocene Humans." *Comptes Rendus Palevol* 16 (2017): 680–89.

Stearn, W. T. "The Background of Linnaeus's Contributions to the Nomenclature and Methods of Systematic Biology." *Systemic Zoology* 8 (1959): 4–22.

Straus, Lawrence Guy, Geoffrey A. Clark, Jesus Altuna, and Jesus A. Ortea. "Ice-Age Subsistence in Northern Spain." *Scientific American* 242 (1980): 142–52.

Stringer, Chris. "The Origin and Evolution of *Homo sapiens*." *Philosophical Transactions of the Royal Society B* 371 (2016): 20150237.

——. "The Status of *Homo heidelbergensis* (Schoetensack 1908)." *Evolutionary Anthropology* 21 (2012): 101–7.

Stringer, C., R. Grün, H. Schwarcz, and P. Goldberg. "ESR Dates for the Hominid Burial Site of Es Skhul in Israel." *Nature* 338 (1989): 756–58.

Stringer, Christopher B., Jean-Jacques Hublin, and Bernard Vandermeersch. "The Origins of Anatomically Modern Humans in Western Europe." In *The Origins of Modern Humans: A World Survey of the Fossil Evidence*, ed. Fred H. Smith and Frank Spencer, 51–135. New York: Alan R. Liss, 1984.

Swofford, David L. *Phylogenetic Analysis Using Parsimony (PAUP).* Version 2.4. Champaign, IL: Illinois Natural History Survey, 1985.

Tallavaara, Miikka, Miska Luoto, Natalia Korhonen, Heikki Järvinen, and Heikki Seppä. "Human Population Dynamics in Europe over the Last Glacial Maximum." *Proceedings of the National Academy of Sciences (USA)* 112 (2015): 8232–37.

Tartar, Elise, and Randall White. "The Manufacture of Aurignacian Split-Based Points: An Experimental Challenge." *Journal of Archaeological Science* 40 (2013): 2723–45.

Tejero, José-Miguel. "Spanish Aurignacian Projectile Points: An Example of the First European Paleolithic Hunting Weapons in Osseous Materials." In *Osseous Projectile Weaponry: Towards an Understanding of Pleistocene Cultural Variability*, ed. Michelle C. Langley, 55–69. Dordrecht: Springer, 2016.

Texier, Pierre-Jean, Guillaume Porraz, John Parkington, Jean-Philippe Rigaud, Cedric Poggenpoel, Christopher Miller, Chantal Tribolo, et al. "A Howiesons Poort Tradition of Engraving Ostrich Eggshell Containers Dated to 60,000 Years Ago at Diepkloof Rock Shelter, South Africa." *Proceedings of the National Academy of Sciences* (USA) 107 (2010): 6180–85.

Thibeault, Adrien, and Sébastien Villotte. "Disentangling Cro-Magnon: A Multiproxy Approach to Reassociate Lower Limb Skeletal Remains and to Determine the Biological Profiles of the Adult Individuals." *Journal of Archaeological Science: Reports* 21 (2018): 76–86.

Thieme, Hartmut. "Lower Palaeolithic Hunting Spears from Germany." *Nature* 385 (1997): 807–10.

——. "Lower Palaeolithic Throwing Spears and Other Wooden Implements from Schöningen." In *Hominid Evolution: Lifestyles and Survival Strategies*, ed. Herbert Ullrich, 383–95. Gelsenkirchen, Germany: Archaea, 1999.

Tierney, Jessica E., Jiang Zhu, Jonathan King, Steven B. Malevich, Gregory J. Hakim, and Christopher J. Poulsen. "Glacial Cooling and Climate Sensitivity Revisited." *Nature* 584 (2020): 569–73.

Tostevin, Gilbert. "Social Intimacy, Artefact Visibility, and Acculturation Models of Neanderthal-Modern Human Interaction." In *Rethinking the Human Revolution: New Behavioural and Biological Perspectives on the Origin and Dispersal of Modern Humans*, ed. Paul Mellars, Katie Boyle, Ofer Bar-Yosef, and Chris Stringer, 341–57. Cambridge: MacDonald Institute Monographs, 2007.

Trinkaus, Erik. "An Abundance of Developmental Anomalies and Abnormalities in Pleistocene People." *Proceedings of the National Academy of Sciences (USA)* 115 (2018): 11941–44.

——. "Anatomical Evidence for the Antiquity of Human Footwear Use." *Journal of Archaeological Science* 32 (2005): 1515–26.

——. "Body Length and Mass." In *Early Modern Human Evolution in Central Europe: The People of Dolní Věstonice and Pavlov*, ed. Erik Trinkaus and Jiří Svoboda, 233–41. Oxford: Oxford University Press, 2006.

——. "Neanderthal Limb Proportions and Cold Adaptation." In *Aspects of Human Evolution*, ed. Christopher B. Stringer, 187–224. London: Taylor and Francis, 1981.

Trinkaus, Erik, Simon W. Hillson, Robert G. Franciscus, and Trenton W. Holliday. "Skeletal and Dental Paleopathology." In *Early Modern Human Evolution in Central Europe: The People of Dolní Věstonice and Pavlov*, ed. Erik Trinkaus and Jiří Svoboda, 419–58. Oxford: Oxford University Press, 2006.

Trinkaus, Erik, and Pat Shipman. *The Neandertals: Changing the Image of Mankind*. New York: Knopf, 1992.

Trujillo, Cleber A., Edward S. Rice, Nathan K. Schaefer, Isaac A. Chaim, Emily C. Wheeler, Assael A. Madrigal, Justin Buchanan, et al. "Reintroduction of the Archaic Variant of NOVA1 in Cortical Organoids Alters Neurodevelopment." *Science* 371 (2021): eaax2537.

Verneau, René. *Les Grottes de Grimaldi (Baoussé-Roussé). Anthropologie*. Monaco: Imprimerie de Monaco, 1906.

Vidal, Céline M., Christine S. Lane, Asfawossen Asrat, Dan N. Barfod, Darren F. Mark, Emma L. Tomlinson, Amdemichael Zafu Tadesse, et al. "Age of the Oldest Known *Homo sapiens* from Eastern Africa." *Nature* 601 (2022): 579–83.

Villa, Paola, Luca Pollarolo, Jacopo Conforti, Fabrizio Marra, Cristian Biagioni, Ilaria Degano, Jeannette J. Lucejko, et al. "From Neandertals to Modern Humans: New Data on the Uluzzian." *PLoS ONE* 13 (2018): e0196786.

Villmoare, Brian, William H. Kimbel, Chalachew Seyoum, Christopher J. Campisano, Erin N. DiMaggio, John Rowan, David R. Braun, J. Ramón Arrowsmith, and Kaye E. Reed. "Early *Homo* at 2.8 Ma from Ledi-Geraru, Afar, Ethiopia." *Science* 347 (2015): 1352–55.

Villotte, Sébastien, and Dominique Henry-Gambier. "The Rediscovery of Two Upper Palaeolithic Skeletons from Baousso da Torre Cave (Liguria-Italy)." *American Journal of Physical Anthropology* 141 (2010): 3–6.

Vlček, Emanuel. *Die Mammutjäger von Dolní Věstonice*. Liestal, Switzerland: Archäologie und Museum 22.

Welker, Frido, Mateja Hajdinjak, Sahra Talamo, Klervia Jaouen, Michael Dannemann, Francine David, Michèle Juliene, et al. "Palaeoproteomic Evidence Identifies Archaic Hominins Associated with the Châtelperronian at the Grotte du Renne." *Proceedings of the National Academy of Sciences (USA)* 113 (2016): 11162–67.

White, E. B., and Katharine S. White. *A Subtreasury of American Humor.* New York: Coward-McCann, 1941.

White, Randall. "Rethinking the Middle/Upper Paleolithic Transition." *Current Anthropology* 23 (1982): 169–92.

White, Tim D., Berhane Asfaw, David DeGusta, Henry Gilbert, Gary D. Richards, Gen Suwa, and F. Clark Howell. "Pleistocene *Homo sapiens* from Middle Awash, Ethiopia." *Nature* 423 (2003): 742–47.

Wilde, Sandra, Adrian Timpson, Karola Kirsanow, Elke Kaiser, Manfred Kayser, Martina Unterländer, Nina Hollfelder, et al. "Direct Evidence for Positive Selection of Skin, Hair, and Eye Pigmentation in Europeans During the Last 5,000 Years." *Proceedings of the National Academy of Sciences (USA)* 111 (2014): 4832–37.

Wissing, Christoph, Hélène Rougier, Chris Baumann, Alexander Comeyne, Isabelle Crevecoeur, Dorothée G. Drucker, Sabine Gaudzinski-Windheuser, et al. "Stable Isotopes Reveal Patterns of Diet and Mobility in the Last Neandertals and First Modern Humans in Europe." *Nature Scientific Reports* 9 (2019): 4433.

Wolpoff, Milford H. "Describing Anatomically Modern *Homo sapiens*. A Distinction Without a Definable Difference." In *Fossil Man—New Facts, New Ideas. Papers in Honor of Jan Jelínek's Life Anniversary,* ed. V. V. Novotný and A. Mizerová, 41–53. Brno, Czech Republic: Anthropos, 1986.

Wolpoff, Milford H., Xinzhi Wu, and Alan G. Thorne. "Modern *Homo sapiens* Origins: A General Theory of Hominid Evolution Involving the Fossil Evidence from East Asia." In *The Origins of Modern Humans: A World Survey of the Fossil Evidence,* ed. Fred H. Smith and Frank Spencer, 411–83. New York: Alan R. Liss, 1984.

Wood, R. E., A. Arrizabalaga, M. Camps, S. Fallon, M.-J. Iriarte-Chiapusso, R. Jones, J. Maroto, et al. "The Chronology of the Earliest Upper Palaeolithic in Northern Iberia: New Insights from L'Arbreda, Labeko Koba and La Viña." *Journal of Human Evolution* 69 (2014): 91–109.

Wood, Rachel E., Cecilio Barroso, Miguel Caparrós, Jesús F. Jordá, Bertila Galván, and Thomas F. G. Higham. "Radiocarbon Dating Casts Doubt on the Late Chronology of the Middle to Upper Palaeolithic Transition in Southern Iberia." *Proceedings of the National Academy of Sciences (USA)* 110 (2013): 2781–86.

Wood, Rachel, Federico Bernaldo de Quirós, José-Manuel Maíllo-Fernández, José-Miguel Tejero, Ana Neira, and Thomas Higham. "El Castillo (Cantabria, Northern Iberia) and the Transitional Aurignacian: Using Radiocarbon Dating to Assess Site Taphonomy." *Quaternary International* 474 (2018): 56–70.

Wragg Sykes, Rebecca. *Kindred: Neanderthal Life, Love, Death and Art.* New York: Bloomsbury Sigma, 2020.

Zhu, Zhaoyu, Robin Dennell, Weiwen Huang, Yi Wu, Shifan Qiu, Shixia Yang, Zhiguo Rao, et al. "Hominin Occupation of the Chinese Loess Plateau Since About 2.1 Million Years Ago." *Nature* 559 (2018): 608–12.

Zilhão, João. "Chronostratigraphy of the Middle-to-Upper Paleolithic Transition in the Iberian Peninsula." *Pyrenae* 37 (2006): 7–84.

——. "Comment to Henshilwood and Marean." *Current Anthropology* 44 (2003): 642–43.

——. "Szeletian, Not Aurignacian: A Review of the Chronology and Cultural Associations of the Vindija G1 Neandertals." In *Sourcebook of Paleolithic Transitions,* ed. Marta Camps and Parth Chauhan, 407–26. New York, Springer, 2009.

Zilhão, João, Diego E. Angelucci, Ernestina Badal-García, Francesco D'Errico, Floréal Daniel, Laure Dayet, Katerina Douka, et al. "Symbolic Use of Marine Shells and Mineral Pigments by Iberian Neandertals." *Proceedings of the National Academy of Sciences (USA)* 107 (2010): 1023–28.

Zilhão, João, William E. Banks, Francesco D'Errico, and Patrizia Gioia. "Analysis of Site Formation and Assemblage Integrity Does Not Support Attribution of the Uluzzian to Modern Humans at Grotta del Cavallo." *PLoS ONE* 10 (2015): e0131181.

Zilhão, João, and Francesco D'Errico. "An Aurignacian Garden of Eden in Southern Germany? An Alternative Interpretation of the geissenklösterle and a Critique of the Kulturpumpe Model." *Paleo* 15 (2003): 1–28.

Zilhão, João, Francesco D'Errico, Jean-Guillaume Bordes, Arnaud Lenoble, Jean-Pierre Texier, and Jean-Philippe Rigaud. "Analysis of Aurignacian Interstratification at the Châtelperronian-Type Site and Implications for the Behavioral Modernity of Neandertals." *Proceedings of the National Academy of Sciences (USA)* 103 (2006): 12643–48.

——. "Grotte des Fées (Châtelperron): History of Research, Stratigraphy, Dating, and Archaeology of the Châtelperronian Type-Site." *PaleoAnthropology* (2008): 1–42.

Zilhão, João, and Paul Pettitt. "On the New Dates for Gorham's Cave and the Late Survival of Iberian Neanderthals." *Before Farming* 3 (2006): 1–9.

Zubrow, Ezra. "The Demographic Modeling of Neanderthal Extinction." In *The Human Revolution: Behavioural and Biological Perspectives on the Origins of Modern Humans*, ed. Paul Mellars and Chris Stringer, 212–31. Edinburgh: University of Edinburgh Press, 1989.

Index

Page numbers in *italics* indicate figures or tables.

Printed and bound by CPI Group (UK) Ltd, Croydon, CR0 4YY

13/06/2023

03226430-0001